READING AND INTERPRETING DIAGRAMS IN AIR CONDITIONING AND REFRIGERATION

READING AND INTERPRETING DIAGRAMS IN AIR CONDITIONING AND REFRIGERATION

by Edward F. Mahoney

A RESTON BOOK
PRENTICE-HALL, INC., Englewood Cliffs, New Jersey 07632

Library of Congress Cataloging in Publication Data

Mahoney, Edward F.
Reading and interpreting diagrams in air conditioning and refrigeration.

Includes index.
1. Air conditioning—Equipment and supplies—
Drawings. 2. Refrigeration and refrigerating machinery
—Drawings. I. Title.
TH7687.5.M33 1983 697.9′3 82-25045
ISBN 0-8359-6483-3
ISBN 0-8359-6483-5 (pbk.)

A Division of Simon & Schuster
Englewood Cliffs, New Jersey 07632

10 9 8 7 6 5 4

Printed in the United States of America

DEDICATION

This book is dedicated to a nameless electrical lineman who, years ago (in the small town of Carman—outside Schenectady, New York), gave the little kid a partial roll of friction tape. He became my hero and started my interest in electricity. I now often wonder whether and how my words and actions—without any special intent on my part—may affect the life of another.

TABLE OF CONTENTS

PREFACE

Reading and Interpreting Diagrams in Air Conditioning and Refrigeration is intended as a companion text for my earlier book, *Electricity for Air Conditioning and Refrigeration Technicians*. In this book I further develop material on the diagram systems used in the trade area.

The topics are presented in a sequence considered practical for course work. It seems reasonable to begin with a review of electricity, a unit on symbols, and a general unit on diagrams and schematics. The remaining topics might well be covered in a sequence other than that followed here.

Each circuit described in this book should be taken only as a possible system for the connection of components. There has been no intention to present any manufacturer's design of circuits. When a technician is working on actual equipment, the manufacturer's drawings should always be consulted.

A compilation of the many symbols used in air-conditioning and refrigeration work is provided in Appendix A. It is suggested that the technician become familiar with every symbol. Other appendixes include power and wiring requirements.

I would like to express my appreciation to Boyce Dwiggins, chairman of the Trade and Technical Dertment, Sheridan Vocational Center, Hollywood, Florida, for his suggestions on the topical material for this book. Finally, I wish to thank my wife, Gloria, for spelling and punctuation corrections and for her excellent work in typing the manuscript.

READING AND INTERPRETING DIAGRAMS IN AIR CONDITIONING AND REFRIGERATION

Chapter 1

BASIC ELECTRICITY: A REVIEW

The interpretation of diagrams is an important part of the air-conditioning and refrigeration technician's job. In order to interpret diagrams properly, a good knowledge of basic electricity is needed. This unit fairly well covers the information about basic electricity needed by the air-conditioning technician. This review should be sufficient for technicians previously trained in basic electricity. The student not yet trained in it should take a more comprehensive approach. Such an approach is beyond the scope of this text, however. A bibliography for the study of related material is included at the end of this book. Read through Chapter 1 either as a review or to obtain an indication of what you should be familiar with.

VOLTAGE, CURRENT, AND RESISTANCE

Voltage, current, and resistance are the three main factors in basic electricity. An understanding of the relationship among the three is the basis for an understanding of electricity.

Voltage:

Voltage, which is measured in volts (V), is the pressure that exists in an electric circuit and causes current to flow. Voltage always exists between two points. For example, it is not proper to say that the voltage at point *A* is 115 volts. It is proper to say that the voltage between point *A* and the ground, for example, is 115 V. The symbol for voltge is "E."

Current:

Current is the flow of electrons. Current is measured in amperes (amps). The symbol for current is "I."

Resistance:

Resistance is the property of the circuit that limits or holds back the flow of current. Resistance is measured in ohms (Ω). The symbol for resistance is "R."

Resistance is dependent on the object's area and length and the material from which the object is made. Copper, for example, is a good conductor of electricity. Copper has a low resistance for a given size and length. Iron wire has a much higher resistance than copper wire of the same size and length. Both copper and iron are considered to be conductors of electricity.

Insulators:

Some materials do not conduct electricity very well. These materials are called insulators. An insulator is nothing more than a material that has extremely high resistance. Insulating material is used to surround wires carrying current. The wire used to wind the coils in motors and contactors is covered with an insulating varnish. Insulating materials are used to keep electricity from flowing in places where it is not wanted. Figure 1–1 shows a compressor whose start, run, and common terminals are insulated from each other and from the metal case by the glass insert.

MAGNETISM

A magnet is a metal or a mixture of metal and some other material that will attract iron, steel, nickel, cobalt, and alloys made with these metals. A magnet has a field around it made up of lines of force. It is the magnetic field that provides the magnet with the ability to attract other metals.

If a bar magnet is held by a string and balanced, it will align itself with the earth's magnetic field. The end of the magnet that is pointed north is the north-seeking pole, and the end of the magnet that points south is the south-seeking pole. The poles are marked N and S, respectively.

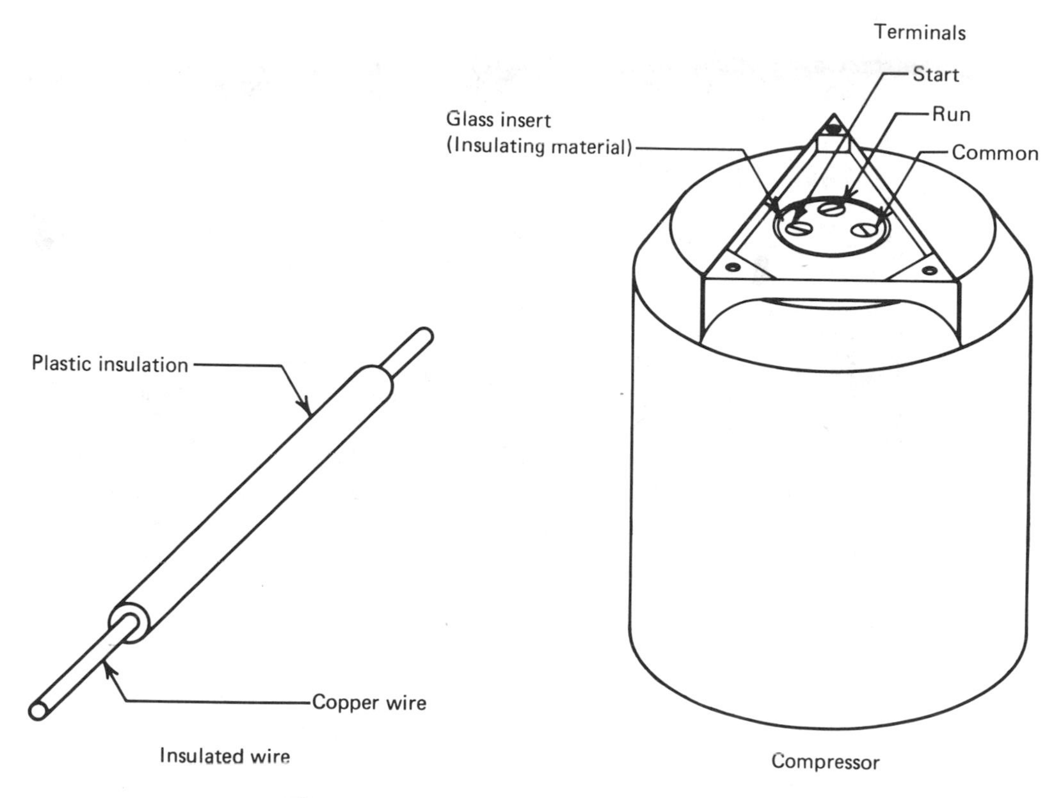

Figure 1–1. Use of insulating material.

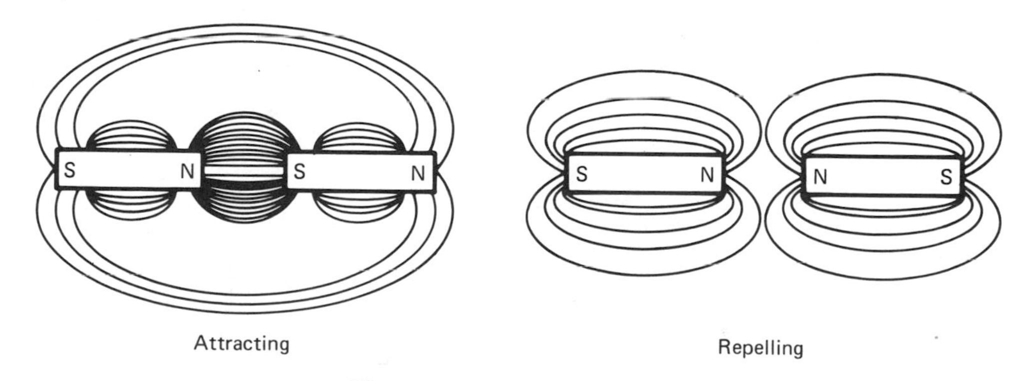

Figure 1–2. Bar Magnets

There is a relationship between the poles of magnets. Unlike poles will attract each other; like poles will repel each other. Figure 1–2 shows the magnetic fields when the poles are attracting and repelling.

ELECTROMAGNETS

Whenever an electric current flows in a wire, the wire will have a magnetic field around it. This is shown in Figure 1–3. When the wire is formed into a coil, the magnetic field is concentrated in the center, as shown in Figure 1–4. The polarity of the electromagnet may be determined by using the lefthand rule. Grasp the coil with the left hand with the fingers extending around the coil in the direction of current flow. The thumb will point to the north pole, as shown in Figure 1–5.

If the core of the coil is a magnetic material such as iron, the magnet will be much stronger than if a nonmagnetic material is used as the core.

Magnetic principles are used in the design of relays, contactors, solenoids, and motors.

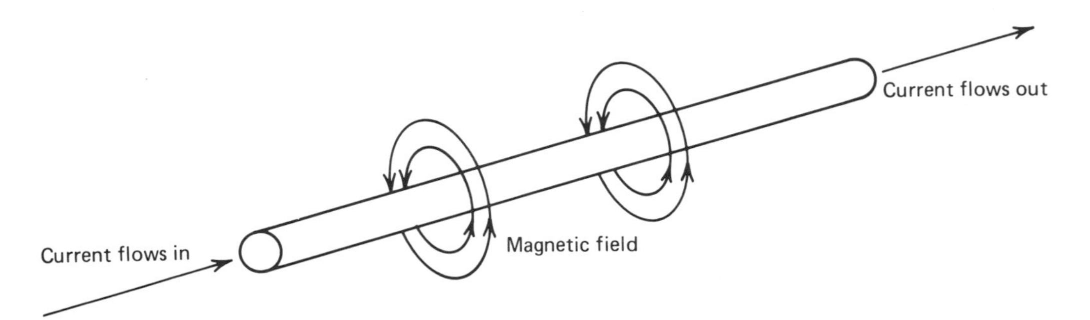

Figure 1–3. Magnetic field around a wire carrying current.

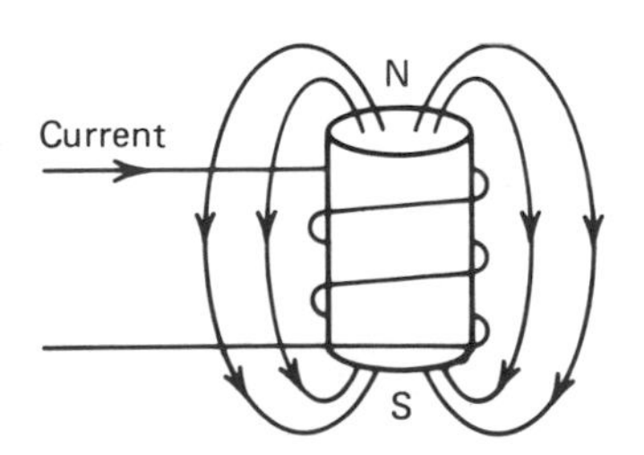

Figure 1–4. Field around a coil of wire.

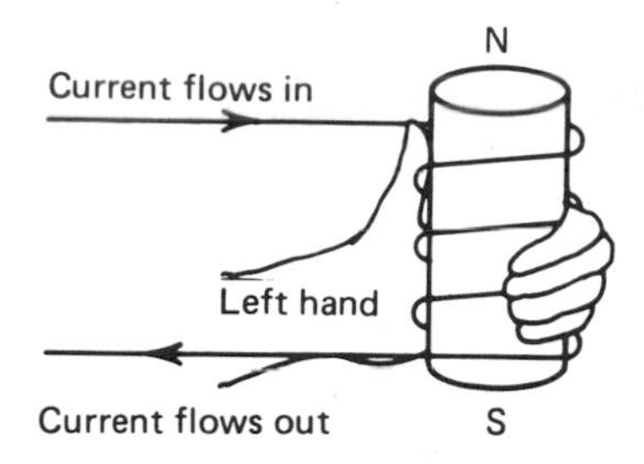

Figure 1–5. Left-hand rule.

OHM'S LAW

Ohm's law states the mathematical relationship between current, voltage, and resistance in electric circuits: *The current in an electric circuit is directly proportional to the voltage and inversely proportional to the resistance.* This relationship is also stated by the following formula: *The current is equal to the voltage divided by the resistance.*

When a formula contains three factors, two of the factors must be known in order to solve for the third. The Ohm's law formula may be stated in three forms, solving for voltage (E), current (I), or resistance (R). The formulas are shown in Figure 1–6.

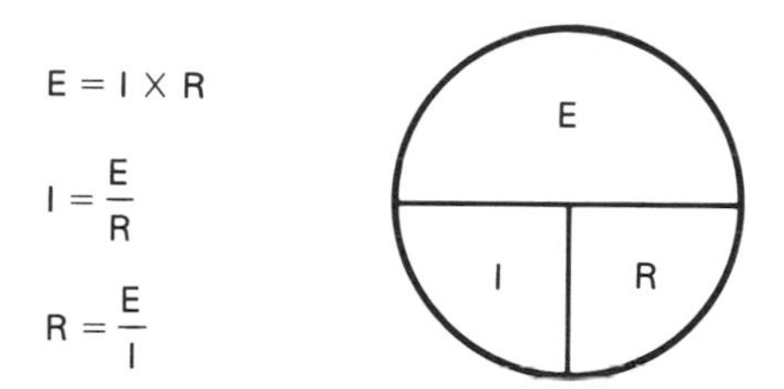

Cover the unknown; the resulting pair will fit one of the formula arrangements.

Figure 1–6. Ohm's law formula.

Ohm's law may be used in solving problems relating to the whole circuit or in solving problems relating to a portion of a circuit.

SERIES CIRCUITS

A series circuit is one in which the components are connected in a sequence such that there is but one path for current to flow in. A series circuit is shown

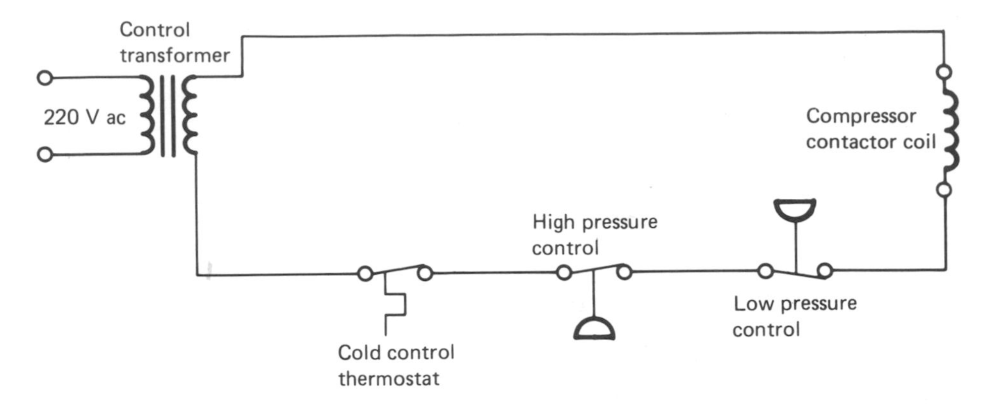

Figure 1–7. Series circuit.

in Figure 1–7. There is only one current path in Figure 1–7, and the same current must pass through each component in the circuit. The purpose of the circuit shown in Figure 1–7 is to provide control of the compressor by means of the compressor contactor. If any of the devices connected in series with the contactor coil should open, the path of the current to the contactor coil would be interrupted and the contactor would drop out, thus removing power from the compressor motor. Most control devices in air-conditioning and refrigeration circuits are connected in series with the device they are controlling.

When components such as resistors are connected in series, a relationship exists between the voltage across the individual components and the supply voltage. A relationship also exists between the individual resistors and the total resistance of the circuit. These relationships are considered to be the laws of series circuits and may be stated as follows:

1. *The total voltage in a series circuit is equal to the sum of the voltages across the individual resistors.*
2. *The total resistance of a series circuit is equal to the sum of the resistances of the individual resistors.*
3. *The same current flows through each component in a series circuit.*

PARALLEL CIRCUITS

A parallel circuit is one in which the same voltage appears across each component in the circuit. It is important to note that the statement specifies the same voltage, not the same amount of voltage. In a parallel circuit, it is actually the same voltage appearing across the components. In Figure 1–8, the compressor motor and evaporator fan motor of an air conditioner are shown connected in parallel. The contacts controlling the circuit are not

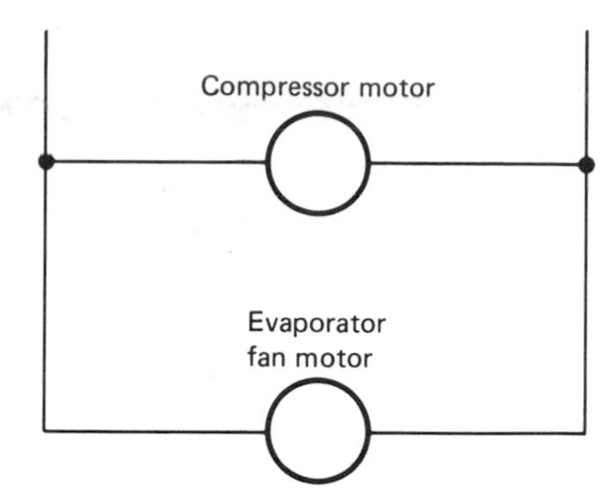

Figure 1–8. Parallel circuit.

included in Figure 1–8 in order that the parallel connection might be seen more clearly.

CURRENTS IN PARALLEL CIRCUITS

In a parallel circuit, the line current is equal to the sum of the currents in the branches. If the compressor motor in Figure 1–8 were drawing 10 amperes and the evaporator fan motor were drawing 3 amperes, then the line current would be the sum of the two currents or 13 amperes.

KIRCHHOFF'S LAWS

Two laws relating to electric circuits sometimes prove useful in problem solving. Stated briefly, they are:

1. *The sum of the voltages around any complete loop is equal to zero.*
2. *The sum of the currents entering a junction is equal to the sum of the currents leaving the junction.*

Returning

In Figure 1–9, two lamps are connected in series across the supply voltage and a resistor is connected in parallel with the lamps. Kirchhoff's first law states that the sum of the voltages around any complete loop is equal to zero. Starting at point *A* in Figure 1–9, take as the polarity of the voltage the first sign encountered. At L1, the voltage is +6 volts; at L2, the voltage is +6 volts; at R1, the voltage is −12 volts. The sum of +6, +6, and −12 is 0 volts. Kirchhoff's voltage law is correct.

Kirchhoff's second law is illustrated by Figure 1–10, which shows the currents traveling in a parallel circuit. There is an indication of 14 amperes

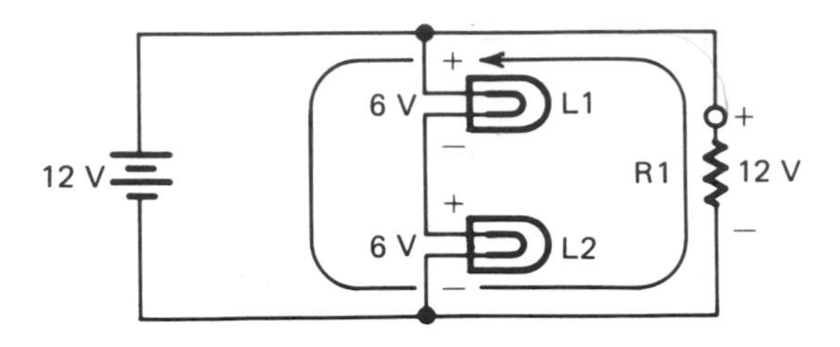

The sum of the voltages
around any complete loop
is equal to zero.

+6 V + 6 V − 12 V = 0

Figure 1–9. Series parallel circuit illustrating Kirchhoff's voltage law.

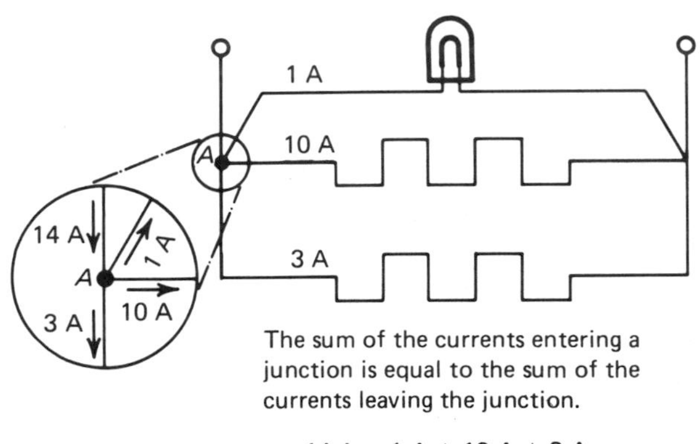

Figure 1–10. Parallel circuit illustrating Kirchhoff's current law.

being fed into point *A*. The currents leaving point *A* are 1 ampere, 10 amperes, and 3 amperes. The sum of 1 ampere, 10 amperes, and 3 amperes is 14 amperes leaving point *A*. Kirchhoff's current law is correct.

In general, Kirchhoff's laws are used in the mathematical solution of complex electrical problems. The laws sometimes prove useful in the solution of more simple problems as well.

POWER AND ENERGY

Power is the rate of doing work. In electrical systems, power is equal to current times voltage, or $P = I \times E$. This formula may be rearranged to solve for power when only the current and resistance are known or when only the voltage and resistance are known.

According to Ohm's law $E = I \times R$. Substituting $(I \times R)$ for E in the formula we get $P = I \times I \times R$ or $P = I^2 R$. Ohm's law also indicates that $I = E/R$. Substituting E/R for I in the power formula $P = I \times E$, we get $P = (E \times E)/R$, or $P = E^2/R$.

Power in electrical systems is measured in watts (W). According to the standard unit of measurement, 746 watts equals one horsepower.

Energy is equal to power times time. Electrical energy is measured in watts times time—usually watt-hours or, for larger values, kilowatt-hours.

Since power is the rate of doing work, or work divided by time, and since energy equals power times time. When these relationships are combined mathematically, we get

$$\frac{\text{Work}}{\text{Time}} \times \text{Time} = \text{Work}$$

The power company charges for kilowatt-hours or power times time, which is essentially a measurement of the amount of work the power company does.

ALTERNATING CURRENT AND VOLTAGE

The most common form of generated electrical power today is alternating current (AC). Alternating current and voltage is the natural way voltages are generated when coils of wire are rotated in a magnetic field. In order to produce direct current (DC), the naturally generated AC must be rectified. Rectification is an extra process. Therefore, AC is more easily produced than DC.

An AC voltage is generated in the form of a sine wave. A sine wave is shown in Figure 1–11. A sine wave of voltage or current consists of 360°,

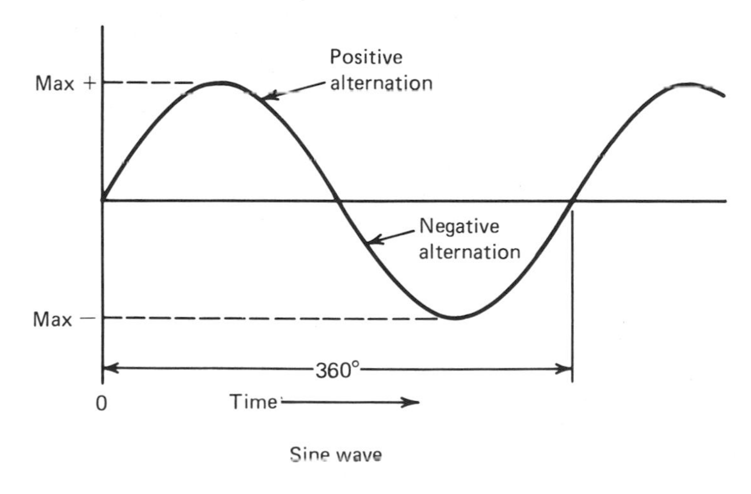

Figure 1–11. Sine wave.

and then the wave repeats itself. The sine wave starts at zero, increases to some maximum positive value, returns to zero, and then goes through zero to a maximum negative value. After reaching the maximum negative value, it again returns to zero. The wave is then ready to repeat itself.

Voltage in the United States is generated at a standard rate of 60 Hertz (Hz). The sine wave repeats itself 60 times each second. The number of sine waves produced each second is called the frequency. Therefore, the frequency of the voltage produced in the United States is 60 Hz. In other parts of the world, voltages may be produced at different frequencies. The only other common frequency in use today is 50 Hz. The voltage and frequency table at the end of this book (Appendix B) lists the standard voltage levels and frequencies generated in many countries.

There is a fixed relationship between the peak value of voltage in a sine wave and the amount of heat or power that will be generated when the sine wave of voltage is applied to a resistor. This relationship or level is called the effective value of voltage. This effective value of voltage is less than the peak value. A comparison is made of the power developed in a DC circuit and in an AC circuit in Figures 1–12 and 1–13.

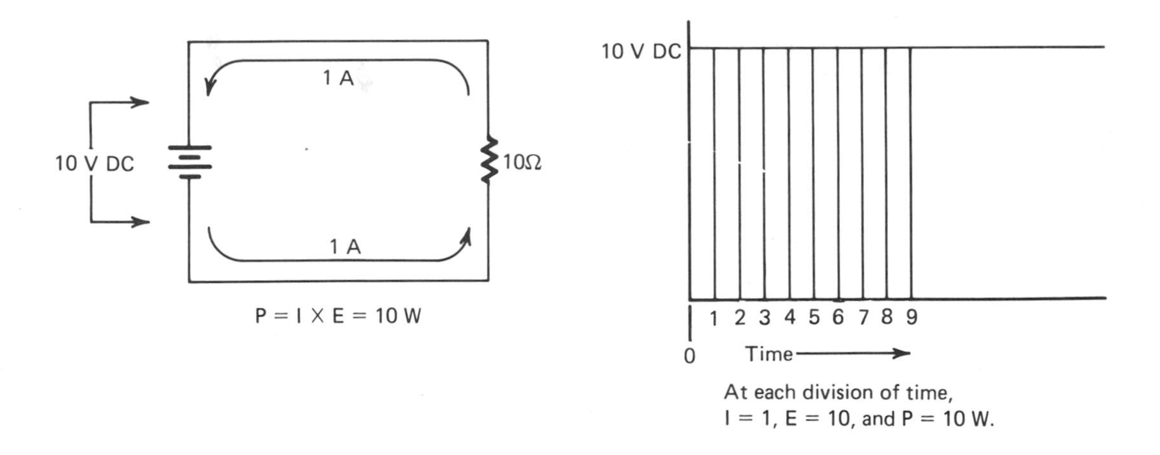

Figure 1–12. Power in a DC circuit.

Figure 1–12 indicates that when 10 volts DC is applied to a 10-ohm resistor, the power developed at each division of time is 10 watts, and the resultant power is therefore 10 watts. The situation is somewhat different when AC is applied to the same resistor. Figure 1–13 shows 10 volts peak AC being applied to a 10-ohm resistor. Starting at zero, where the voltage is also zero, time is divided in nine more steps. From zero to the peak, the voltages will be 0, 1.74, 3.42, 5.00, 6.43, 7.66, 8.66, 9.39, 9.85, and 10.00 volts at the peak. To find the power at each of the ten points, the formula $P = E^2/R$ is

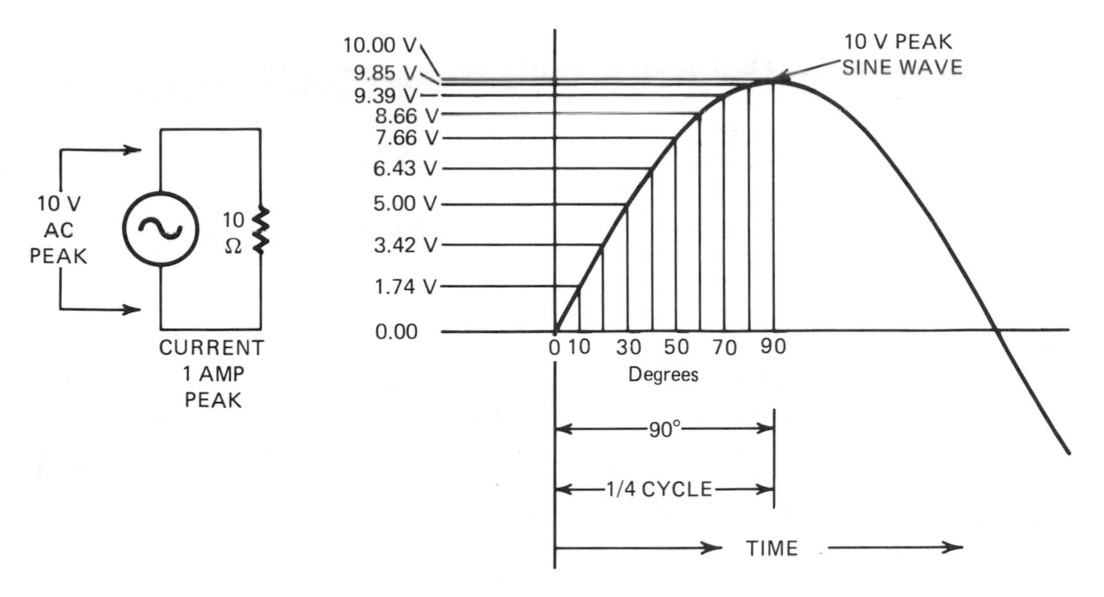

Figure 1–13. Voltage variations of a 10V peak sine wave in 10° steps from 0 to 90°.

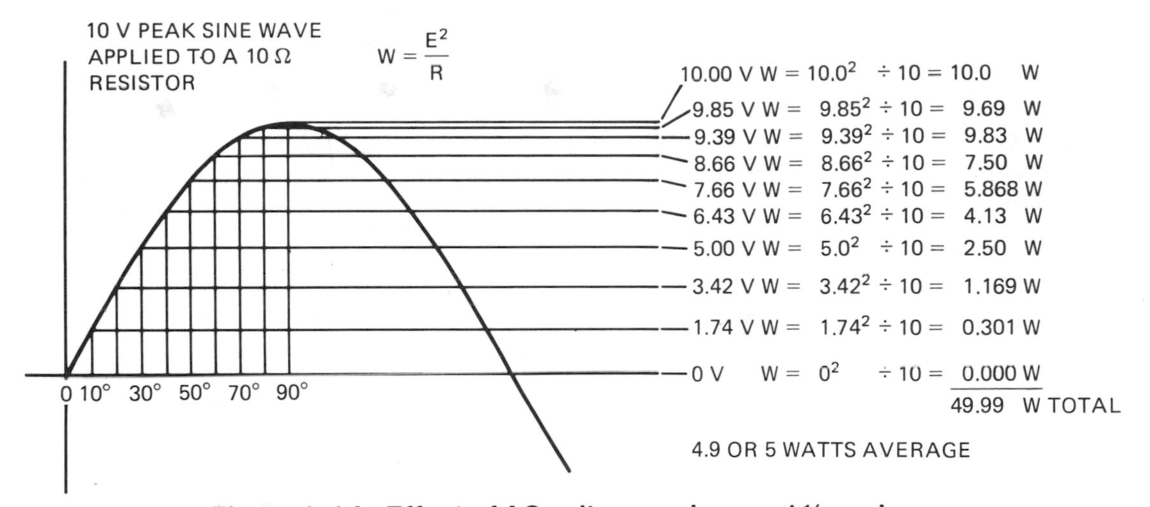

Figure 1–14. Effect of AC voltage and current ¼ cycle.

used. The power at the ten points is 0, 0.301, 1.169, 2.5, 4.13, 5.868, 7.5, 8.83, 9.69, and 10.0 watts at the peak of 10 volts. Figure 1–14 shows the power curve for ¼ cycle of operation. The curve repeats for each ¼ cycle. The average power for the ten points is 4.999 watts or 5 watts (the 50-watt total divided by 10). These three figures show that AC power is just half as effective power at the peak of the wave form.

With the 10-ohm resistor and the peak voltage of 10 volts, the peak current would be 1 ampere. The peak power would be 10 watts. The average power would only be 5 watts, or 0.5 times the peak power. Mathematically, the effective value of the voltage is 0.707 times the peak, and the effective value of the current is 0.707 times the peak. Multiplying the two together results in $0.707 \times 0.707 = .5$.

Standard voltmeters and ammeters have indications on the meter scale of the effective values of voltage or current, but not the peak values.

POLYPHASE ALTERNATING CURRENT

Most of the electrical energy generated in the United States today is in the form of three-phase voltages. Three-phase generators or alternators more effectively use the space within the unit. Three-phase voltages are generated by placing the three-phase coils of the unit 120 electrical degrees apart. The voltages generated in these coils will have a phase relationship of 120°. This relationship is shown in Figure 1–15. In Figure 1–15*b*, phase *B* starts

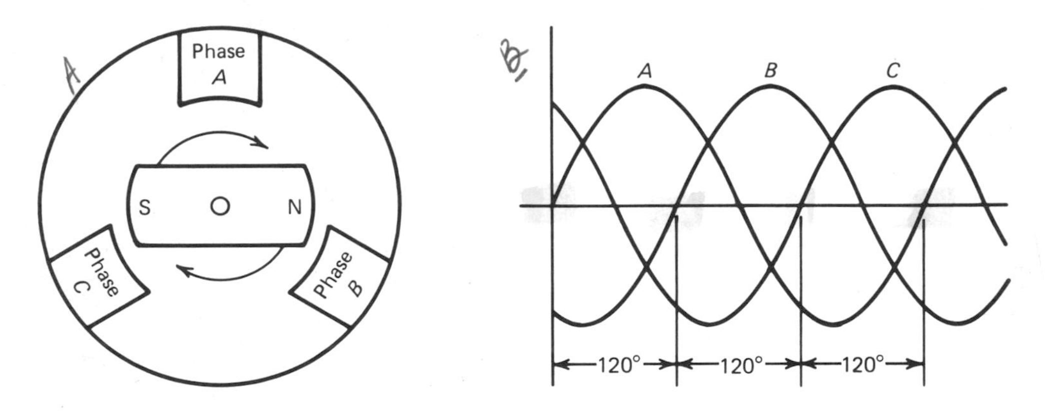

Figure 1–15. Three-phase alternator with output voltages.

positive from zero a distance of 120° from where phase *A* started positive from zero. Phase *C* is shown starting positive from zero a distance of 120° from where phase *B* started positive from zero.

Voltages with the phase relationship shown in Figure 1–15*b* will be generated in the stator coils when the rotor (magnet) of the alternator shown in Figure 1–15*a* is rotated.

POLYPHASE CIRCUIT CONNECTIONS

Three-phase circuits are connected in either wye or delta systems. In the wye system, the coil voltage is equal to the line voltage divided by the square root

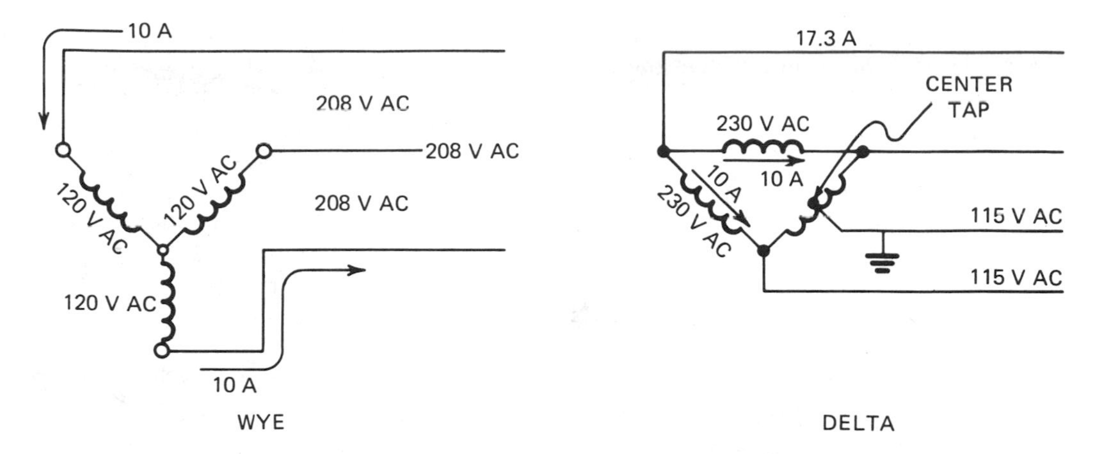

Figure 1–16. Voltage and currents in wye and delta systems.

of 3 (1.732). The coil current in the wye system is equal to the line current. In the delta system, the coil and line voltages are equal. The coil current in the delta system is equal to the line current divided by the square root of 3. Figure 1–16 shows the voltages and currents in the wye and delta systems.

INDUCTANCE

Inductance is the property of an electric circuit that opposes a change in current flow. Inductance exists in an electrical circuit whenever a magnetic field passes through a coil of wire. Whenever a magnetic field passes through a coil of wire, a voltage will be induced in the wire.

Consider the coil of wire in Figure 1–17. An AC voltage is connected to the coil. The AC voltage causes an alternating current to flow through the coil. An AC current is changing constantly. The changing AC current will set

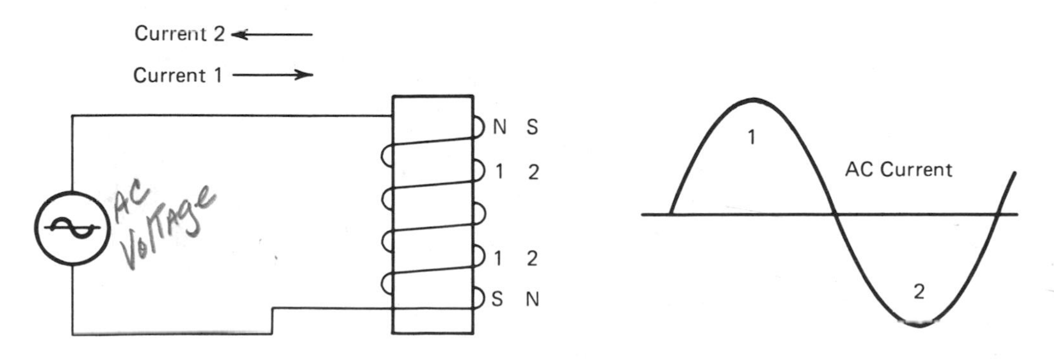

Figure 1–17. Magnetic polarity change with AC current.

up a changing magnetic field around the coil. During the positive alternation (1), the current enters the top of the coil and sets up a north pole at the top and a south pole at the bottom (left-hand rule). During the negative alternation (2), the current enters the bottom of the coil and sets up a north pole at the bottom and a south pole at the top. The continuously changing magnetic field set up around the coil will induce a voltage in the coil opposing the change that produced it. This voltage is called counter electromotive force (CEMF); the more common term is back electromotive force (BEMF).

One other important effect of inductance in a circuit is that it causes the current to lag the voltage. The relation of current to voltage is shown in Figure 1–18.

Any coil of wire carrying AC current exhibits the property of inductance, and the current will lag the voltage by some angle less than 90°. There will always be some resistance in a circuit. Examples of circuits in which

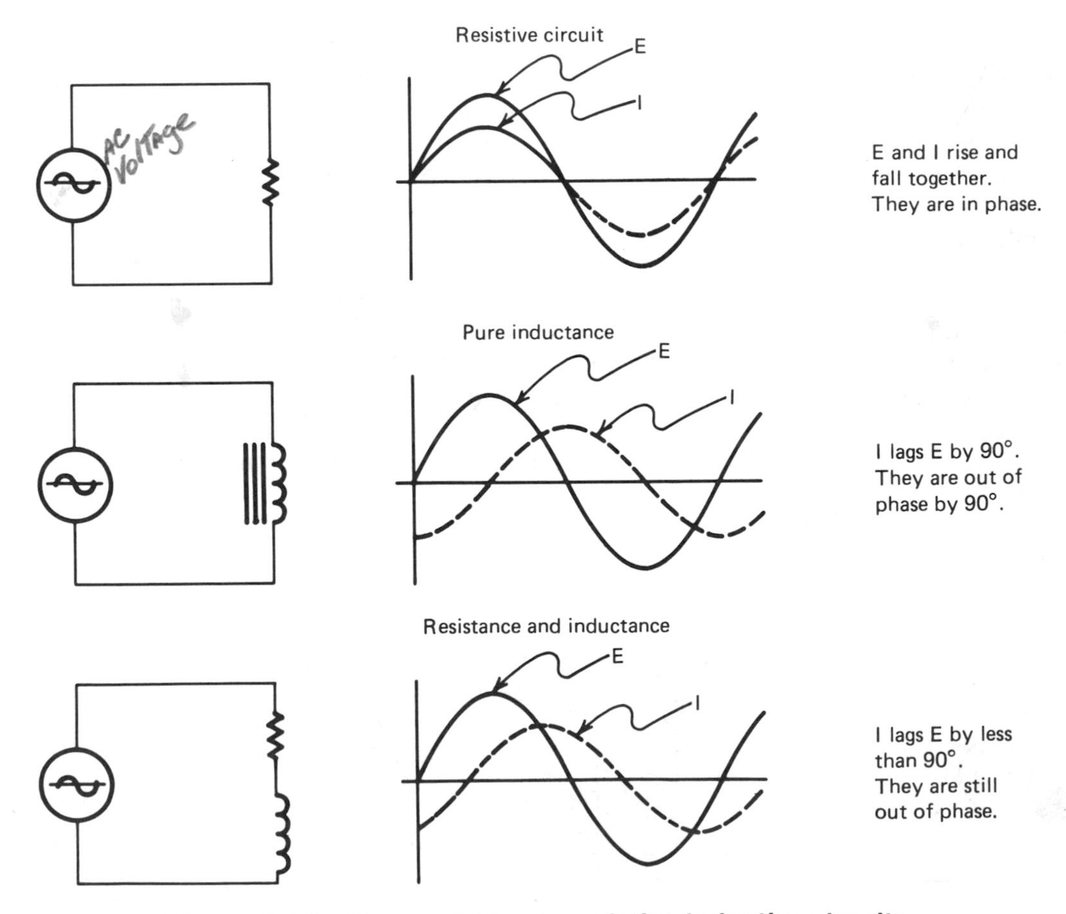

Figure 1–18. Phase relation in resistive-inductive circuits.

inductance exists are contactor or relay coils and solenoid coils. Inductance also exists in most motor circuits.

CAPACITANCE

A capacitor is a device that can store electricity. Technically, a capacitance exists whenever two conductors are separated by an insulator and there is a difference of potential (voltage) between them. Figure 1–19 shows two

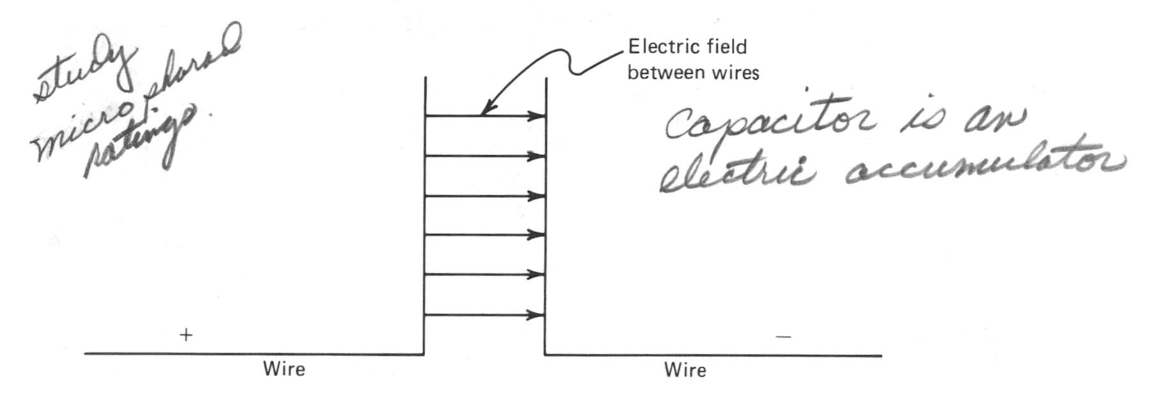

Figure 1–19. Capacitance: two conductors and an insulator.

wires separated by air, an insulator. A small amount of capacitance exists between the two wires. The capacitance can be increased if the area of the capacitor sections is increased; it can be further increased if the plates are moved closer together (see Figure 1–20).

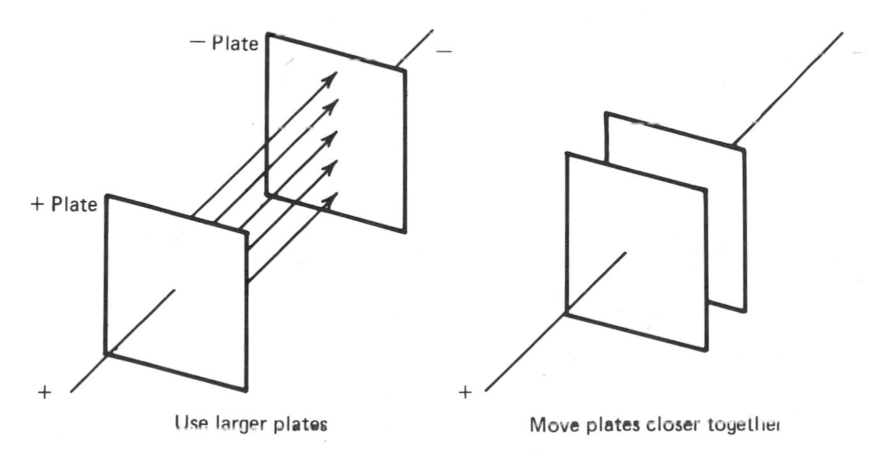

Figure 1–20. Ways to increase capacitance.

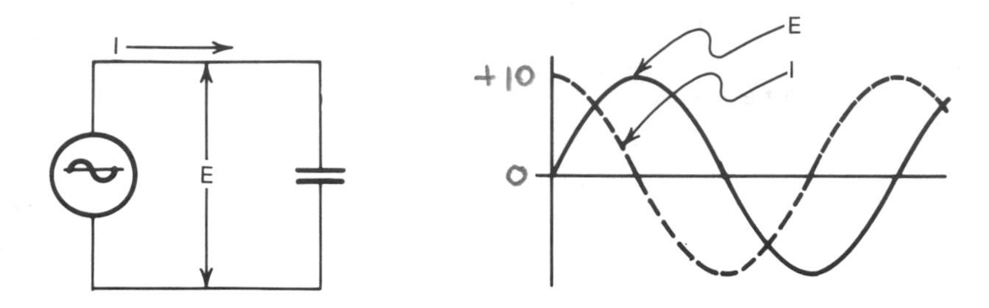

Figure 1–21. Current and voltage in a capacitor.

The property of a capacitor is to oppose a change in voltage. Capacitors are often used in DC power supplies to filter out variations in voltage. In AC circuits, the voltage is always varying. The effect of a capacitor in such circuits is to cause a shift in the phase of the current.

In a capacitor, the current leads the voltage across the capacitor by 90°. This relationship is shown in Figure 1–21. The ability of a capacitor to cause a lead in current phase angle makes the capacitor very useful in motor circuits. The capacitor's leading phase angle provides for a high starting torque in some AC motors because of the large phase shift between the currents in the start and run windings of the motor.

PHASE SHIFT AND POWER FACTOR

In a resistive circuit, the current and voltage rise and fall together, thus ensuring the development of positive power at all times. This relationship is shown in Figure 1–22. Algebraically, when E and I are positive, $E \times I = +$ power, when E and I are both negative, $-E \times -I = +$ power. (Algebraically, $+ \times + = +; + X - = +$).

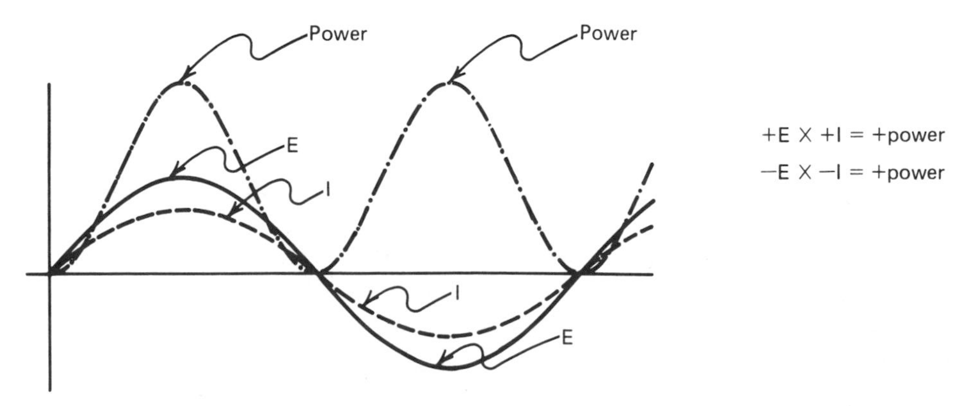

Figure 1–22. Only positive power is developed when current and voltage are in phase.

In an inductive circuit, if the current were lagging the voltage by 90°, then both positive and negative power would be produced. This is shown in Figure 1–23. The negative power is not power used; it is power returned to the power company. If the current and voltage are 90° out of phase, then the positive power sent by the power company is equal to the negative power returned to the power company. No power is used.

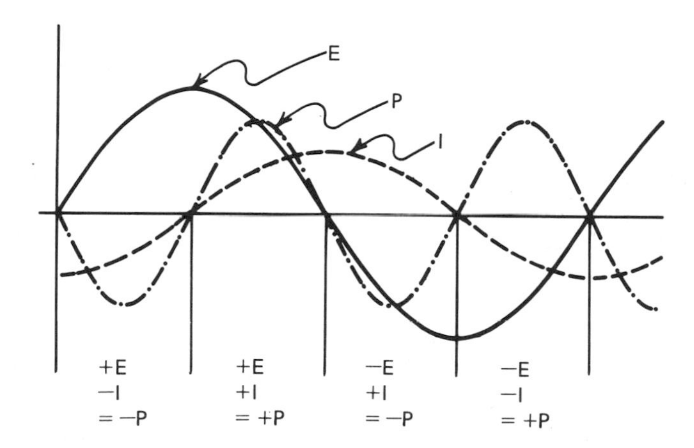

Figure 1–23. Positive and negative power.

In practical cases, current is never 90° out of phase with the voltage. In a motor circuit, the current might lag the voltage by 30°, as shown in Figure 1–24.

The negative power developed in the circuit is small compared with the positive power, but all of the power supplie is not used; some is sent back to the power company. When computing power used, this return of power must be considered. The formula for power in AC circuits is $P = I \times E \times pf$, where pf stands for the power factor. Mathematically, the power factor is the cosine

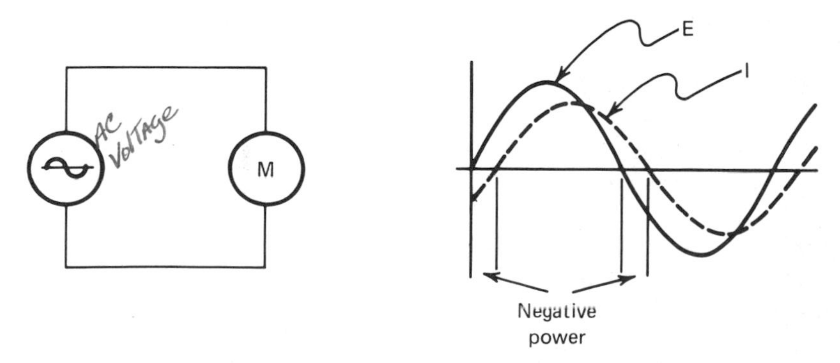

Figure 1–24. Production of negative power in a motor circuit.

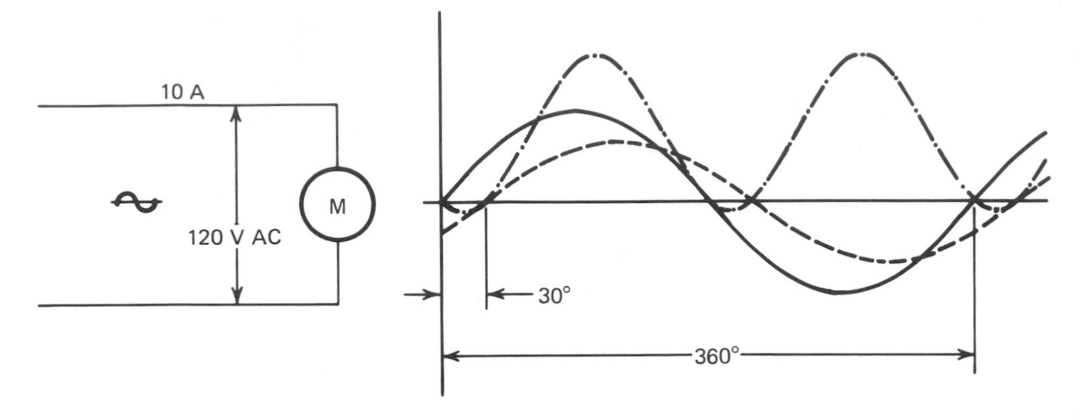

Figure 1–25. Use of power factor.

of the phase angle between the current and the voltage. Power factor use is shown in Figure 1–25. A motor is connected across a 120-volt line. The current drawn by the motor is 10 amperes, lagging the voltage by 30°. An AC motor always draws a lagging current.

$$P = I \times E \times pf$$

$$P = 10 \times 120 \times 0.866 \qquad pf = \cos$$

$$P = 1039 \text{ watts} \qquad \cos 30^\circ = 0.866$$

The power company sent 1200 volt-amperes (VA). The motor used 1039 watts and returned 161 volt-amperes to the power company.

MOTORS

A number of different types of motors are used in air-conditioning and refrigeration circuits. It is important that the technician be familiar with each type of motor, its circuit requirements, and its operating capabilities. The common types of motors being put into use in the trade today are:

1. Shaded pole motor
2. Split capacitor motor
3. Induction motor
4. Capacitor-start motor
5. Capacitor-start, capacitor-run motor
6. Three-phase motor, both wye- and delta-connected

The service technician should make a point of learning the characteristics and connection requirements of each type of motor. The bibliography lists other materials to be used for study.

study Series + parallel circuits + combination

Review Questions

Chapter 1

1. As the resistance in a circuit is increased, the current in the circuit will __increase__ (increase, decrease).
2. Insulators conduct electricity __pooly__ (well, poorly).
3. Glass is a good example of a (an) __insulator__ (insulator, conductor).
4. Unlike poles of a magnet __attract__ (attract, repel) each other.
5. Whenever current passes through a wire, a __magnetic__ __field__ exists around the wire.
6. The __left__ -hand rule is used to determine the polarity of an electromagnet.
7. In order to find the current in a circuit, divide the __voltage__ by the __resistance__ of the circuit.
8. The __same__ current flows through each component of a series circuit.
9. The total voltage around any complete loop is equal to __zero__.
10. The total line current in the parallel circuit is equal to the __sum__ of the currents in the branches.
11. The __sum__ of the currents entering a junction is equal to the __sum__ of the currents leaving the junction.
12. The power company is paid for the amount of __work__ it does.
13. The formula for power in DC circuits is P = __I · E ?__.
14. The formula for power in AC circuits is P = __I · E · PF__.
15. Voltages are produced naturally in an alternator in the form of a ______.
16. The standard line frequency in the United States is __60__ Hertz.
17. The effective value of an AC voltage is __.707__ times the peak value.
18. Standard voltmeters and ammeters indicate the __effective__ (effective, peak) value.

19. In a three-phase system, the voltages are 120 degrees out of phase with each other.
20. The coil voltage is equal to the line voltage in a delta system.
21. The property of inductance is to oppose the change in current flow
22. Current lags the voltage in an inductive circuit.
23. Capacitance may be increased by decreasing (increasing, decreasing) the distance between the plates of a capacitor.
24. The current in a capacitor leads the voltage across the capacitor.

Chapter 2

SCHEMATIC CONTROLS

In order to explain the working relationship of the electrical parts of an air conditioner or refrigerator system, it is necessary to draw a picture of the parts. If the parts are drawn to look like the actual parts, then the internal working mechanisms might not be seen. For this reason, schematic symbols are used to represent some parts. The operating part is presented in a standard form that is accepted as being the indication of that particular part or device by the majority of the workers in the industry or trade area.

Students in the air-conditioning and refrigeration industry must be familiar with the standard symbols used in that industry if they are to interpret properly schematic diagrams of refrigerators and air-conditioning systems. Once the symbols are understood, interpreting the overall systems becomes a simple procedure of following the interconnecting wires.

MANUAL CONTROLS

Manual controls are used in air-conditioning and refrigeration systems to provide direct access to the system's operation. Since the controls are operated manually, no outside activating power source is indicated.

Most symbols have a direct relationship to the device they symbolize. For example, Figure 2–1 shows the schematic symbol of a normally open

Figure 2–1. Push-button switch, normally open.

push-button switch. The symbol consists of the input and output terminals, represented by small circles, and the device that connects the terminals when the button is pressed.

Actually, the schematic symbol is a very close representation of an actual push-button switch. It is obvious that when the button is pressed down, contact will be made between the input and output terminals through the line connecting them.

The operation and symbol of a normally closed-push button switch are as easy to understand as those of the normally open switch. Figure 2–2 shows

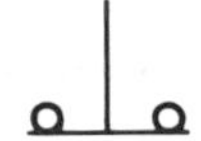

Figure 2–2. Push-button switch, normally closed.

a normally closed push-button switch. As with the normally open push-button switch, the input and output terminals are shown as small circles. Note that the line representing the contacting circuit connects the input and output terminals at the bottom of the terminals. When the button is pressed, the contacting line will move down away from the terminals, breaking the connection between the terminals.

A number of manually operated controls are used in air conditioning and refrigeration. The symbols shown in Figure 2–3 are all examples of controls that are operated manually. The operation of each is fairly clear.

TEMPERATURE CONTROLS

Temperature controls are quite common in air-conditioning and refrigeration systems. The symbol for the temperature-sensing element is shown in Figure 2–4. It is the combination of the temperature-sensing element with switches that provides temperature control. Many different controls are needed, but in each there is a common type of temperature-sensing element. Different temperature controls are shown in Figure 2–5.

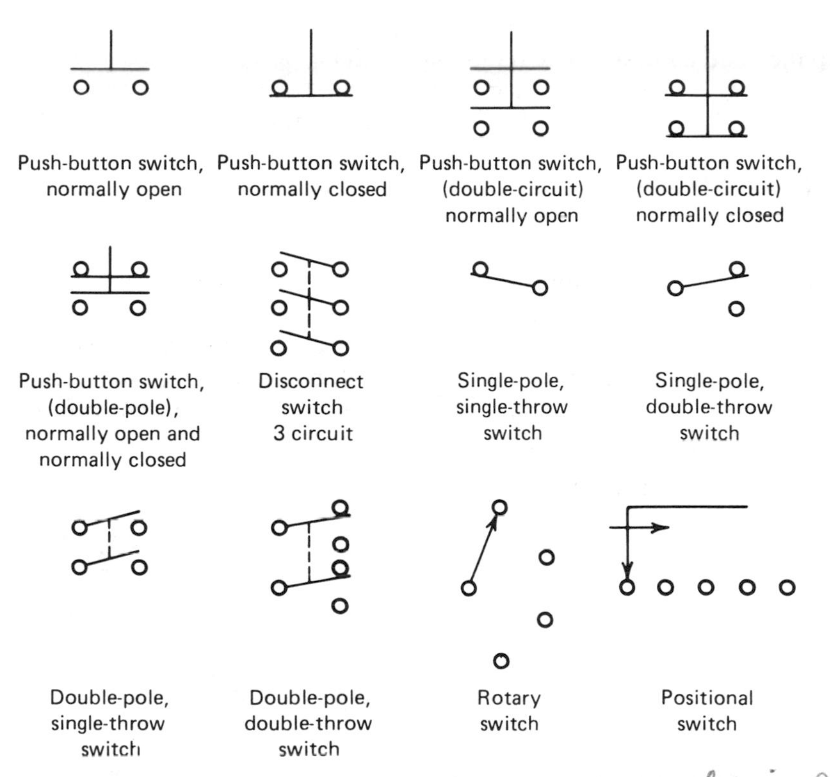

Figure 2–3. Manually operated switches.

Figure 2–4. Temperature-sensing element.

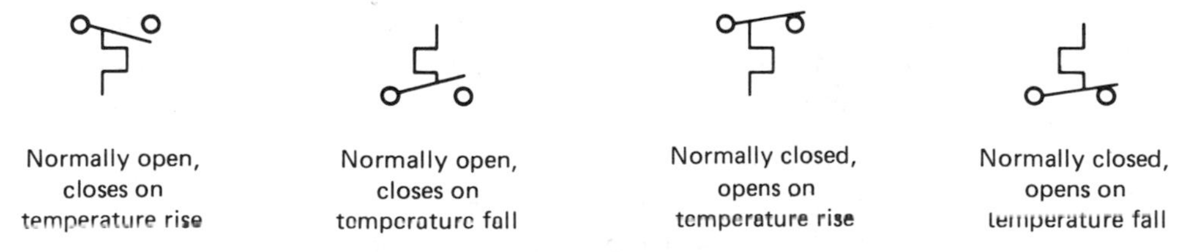

Figure 2–5. Temperature-controlled switches.

The temperature-sensing element should be regarded as a device that expands and contracts with temperature changes. The expansion or contraction of the element opens or closes the switch, depending on the way it is mechanically connected to the movable arm of the switch.

BIMETAL THERMOSTATS

Bimetal thermostats are often used in compressor connections to sense the temperature of the compressor. If the temperature of the compressor is too high, the bimetal strip in the thermostat snaps open, breaking the connection between the contacts and removing electrical power from the compressor motor. The schematic symbol for a thermal, bimetal thermostat is shown in Figure 2–6a. The term Klixon, a brand name, is often used when refering to a thermal, bimetal snap thermostat.

Figure 2–6. Thermal snap switch (KLIXON)

In some systems, it is important to sense the starting current as well as the running temperature of the compressor. In a system of this type, the thermostat shown in Figure 2–6b would be used. Electrical power would be fed to the circuit at terminal 1. The run winding of the motor would be connected to terminal 2. The start winding of the motor would be connected to terminal 3. Should the temperature of the compressor rise above a safe limit, the bimetal strip would snap open just as the unit in Figure 2–6*a* does. The added feature of the unit shown in Figure 2–6*b* is the heater resistor. During compressor start-up, if a start winding current is drawn that is too high or too long, the internal resistor between terminals 2 and 3 of the thermostat will heat up. This excess heat will cause the bimetal disk to snap open, interrupting the connection between terminals 1 and 2.

Another symbol often used to indicate a temperature-sensing element is shown in Figure 2–7. This symbol is used to show an overload device that is temperature activated. An example of this device, a circuit breaker, is shown in Figure 2–8.

Figure 2–7. Thermal overload.

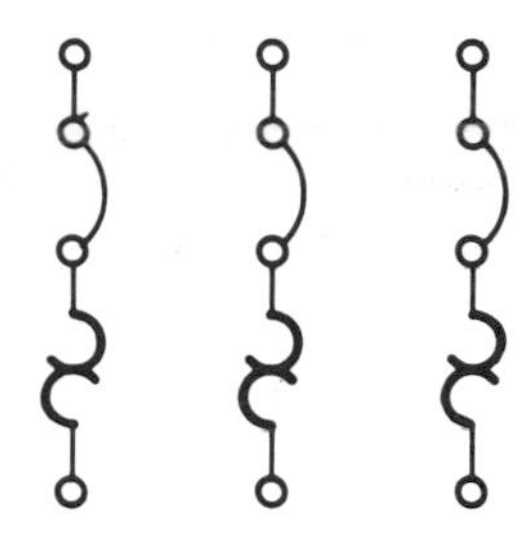

Figure 2–8. Circuit breaker.

PRESSURE CONTROLS

Pressure controls or pressure switches are symbolized by the drawings of the switch and the device that indicates a pressure-sensing element. The pressure sensing element is shown in Figure 2–9.

Figure 2–9. Pressure-sensing element

There are four possible situations that should be indicated in a system's schematic drawing: switch closed, opens on rising pressure; switch closed, opens on falling pressure; switch open, closes on rising pressure; and switch open, closes on falling pressure. The four situations are shown in Figure 2–10. The pressure controls commonly used in compressor circuits are the high pressure control (HPC) and the low pressure control (LPC). The high pressure control switch is normally closed, and opens on a rise in pressure. The low pressure control switch is normally closed, and opens on a fall in pressure.

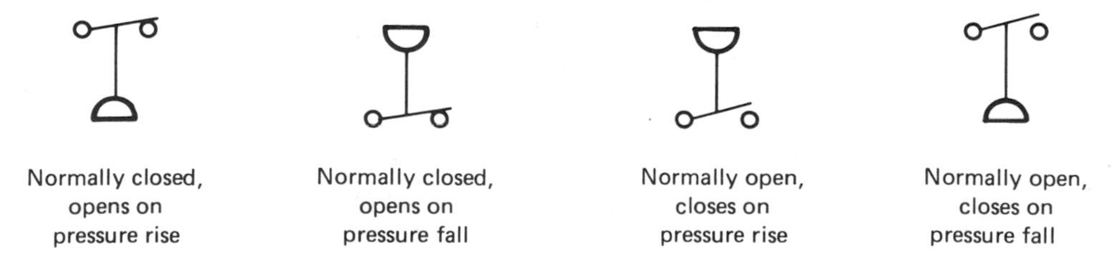

Figure 2–10. Pressure switches.

FLOW CONTROLS

Flow controls or flow switches are used to control the movement of liquids or gases through a system or past a particular point. There are four situations to be indicated. Decreased or increased flow causes a switch to open; decreased or increased flow causes a switch to close. The symbols for these devices are shown in Figure 2–11.

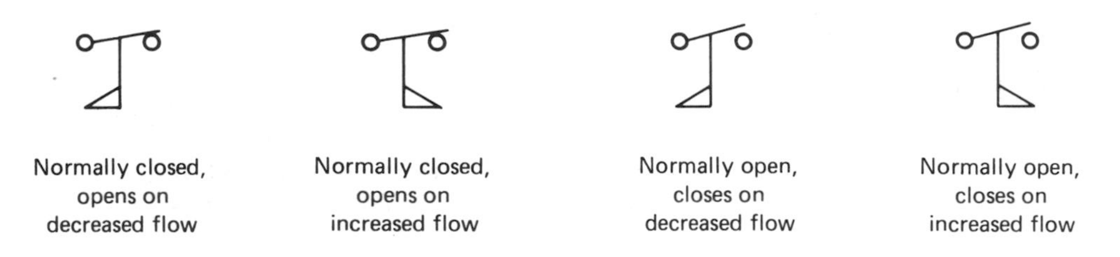

Figure 2–11. Flow switches.

TIMING CONTROLS

Timing controls are similar in operation to pressure controls, temperature controls, and flow controls. The two situations requiring an indication of control action are shown in Figure 2–12. A time delay contact closing and a preset time is shown in Figure 2–12a. A time delay between a preset time and the opening of a switch is shown in Figure 2–12b.

Figure 2–12. Time-delay switch

Two main types of time-delay systems are used in air conditioning and and refrigeration—the bimetal time-delay switch and the clock timer. The bimetal or thermal time-delay switch is often used when a delay of a short period is needed. An example is when starting two or more compressors. To ensure that no more than one compressor comes on the line at a time, a time-delay switch is used to control application of control voltage to the second compressor contactor coil.

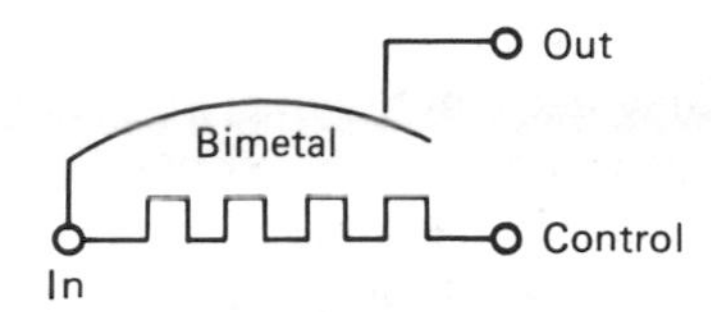

Figure 2–13. Thermal time-delay switch.

The symbol for the bimetal or thermal time-delay switch is shown in Figure 2–13. When electrical power (usually 24 volts AC) is applied between the input and control terminals, a heating of the control resistor results. As the bimetal strip absorbs heat from the resistor, it bends, finally making contact with the output terminal. The 24 volts AC is thereby made available at the output terminal. This occurs some delayed time after application of electrical power to the time-delay relay. The heater will remain hot and the bimetal strip will bend, making contact with the output terminal, as long as 24 volts is available at the input terminal.

The clock timer is another delay timer that is often used in air conditioning and refrigeration. It is most often used when a delay of considerable time is needed. An example is the defrost timer in a refrigeration system. The timing is controlled by a small clock motor that operates a cam. The cam closes a set of contacts for a fixed period. The length of time the contacts are closed is dependent on the cam shape and the motor speed. The symbol for a clock timer is shown in Figure 2–14.

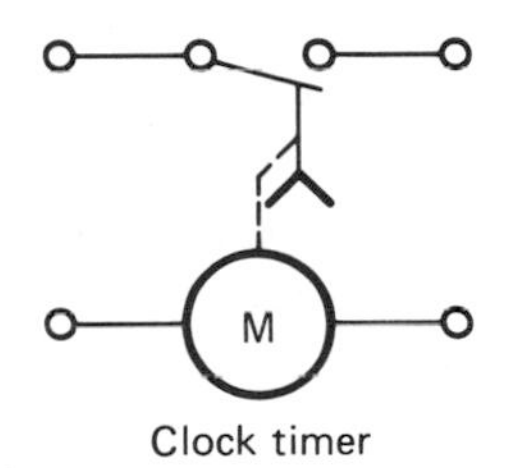

Figure 2–14. Clock timer.

Many different circuits are used in defrost-timing control at the present time. The timer itself is most often just a motor and a cam-operated switch.

ANTI-SHORT-CYCLE CONTROL

An increasing number of electric timers are coming into use in air-conditioning systems. One important application of this timer is in the control of short-cycle operation. Short-cycle operation comes about when line electrical power is interrupted for a short time and then reapplied. If the equipment compressor were in operation when power was interrupted, and it then tried to restart as the power came back on the line, an overcurrent situation would exist. When line power was removed, the compressor would stop. High back pressure would be maintained in the system for a short time. If power were again applied, the compressor would attempt to start against the high back pressure. This would result in extra high starting current.

To overcome this problem, an anti-short-cycle control is often used. This device is nothing more than a time-delay circuit that senses the interruption of power and controls the reapplication of power until after a suitable time delay. Time delays with electronic control are often used in anti-short-cycle controls.

The symbol often seen for electronic anti-short-cycle devices is shown in Figure 2–15. The 24-volt input would be that normally fed to the compressor

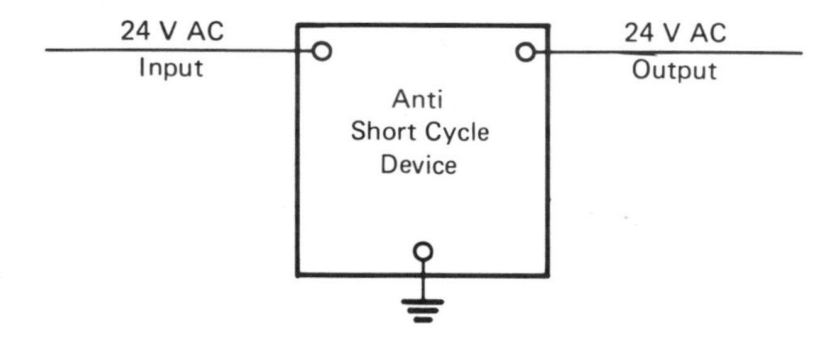

Figure 2–15. Electronic anti-short-cycle device.

contactor coil. During an attempted short-cycle restart, power would be kept from the compressor contactor coil until after a suitable time delay.

An extensive collection of electrical symbols is included in Appendix A. As the control of air-conditioning and refrigeration systems becomes more elaborate, the need to understand use and operation of controls becomes more important. Study the symbols given in this text as well as control manufacturers' parts manuals.

Review Questions

Chapter 2

Instruction: Indicate what each symbol represents and give a description of its operation. See Appendix A.

1.

2.

3.

4.

5.

6.

7.

8. motor

9.

10.

1. N/O start switch
starts a piece of equipment, energizes contacts

2. N/C Flow switch (sail switch)
used to indicate air or water movement.

3. Thermal snap switch
used to indicate temp.

4. N/O time delay

5. N/O float switch (liquid switch)

6. electrical earth ground.

7. conductors connected

8. Capacitor motor

9. Pressure switch

10. thermal overload

Chapter 3

DIAGRAMS

Four main types of drawings are presently being used in the air-conditioning and refrigeration field: pictorial, schematic, wiring, and ladder diagrams. In this chapter, the four drawing systems will be compared. It should be noted that each drawing system has a particular reason for being prepared, and the system should be used for that reason.

PICTORIAL DIAGRAMS

Pictorial diagrams are drawings showing the electrical route from point to point. Each component is drawn as realistically as possible. A light bulb in a refrigerator pictorial diagram is shown looking much like a light bulb. A switch is drawn to resemble a switch, and a fan looks like a fan. Examples of these three devices are shown in Figure 3–1, which is a small section of a refrigerator pictorial diagram.

The wiring used to connect the components in pictorial diagrams is shown in a color code. The wires may be traced from point to point on the diagram as well as in the actual unit. In Figure 3–1, power is fed to the door switch center terminal through the red wire. When the door is closed, contact is made within the switch to the output where the blue wire feeds the power into the fan motor. The blue wire out of the fan connects to the common

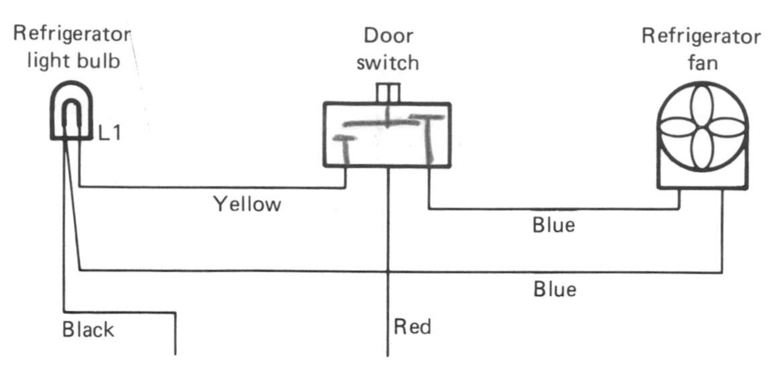

Figure 3–1. Partial pictorial diagram of refrigerator.

terminal of the refrigerator light. The black wire connected to this same terminal provides a connection to the other side of the power source. The fan motor will be running when the refrigerator door is closed. When the door is opened, the door switch breaks the connection between the red input wire and the blue output wire. The switch makes a connection between the red input wire and the yellow output wire. The yellow wire feeds power to the refrigerator light. The black wire connected to the other side of the light is the return electrical connection. when the refrigerator door is opened, the light will be turned on.

Following the circuit in a pictorial diagram is not usually a difficult task. Pictorial diagrams are often used to show refrigerator circuits. In the more complex air-conditioning systems, the schematic, wiring or ladder diagrams are often used.

SCHEMATIC DIAGRAMS

The schematic diagram is a system for showing connections of electrical components in which symbols are used to indicate the components and straight lines are used to indicate the interconnecting wires. In a schematic diagram, almost all interconnecting lines are drawn either horizontally or vertically in order to present a clear and neat descriptive drawing of a circuit.

In a schematic diagram, switches, relays, and contactors are shown in the de-energized position. Standard procedure requires that all drawing symbols show the de-energized position.

Figure 3–2 is the schematic diagram representation of Figure 3–1, the section of the refrigerator pictorial diagram. The schematic diagram shows only the relationship of the components. The actual point of contact for circuit connections cannot be determined from the schematic diagram. For example, the pictorial diagram, Figure 3–1, shows that the (blue) return wire from the fan motor is connected at the common connection (black) terminal

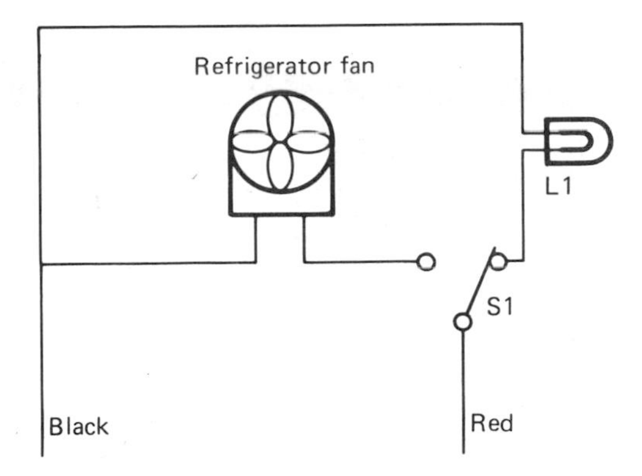

Figure 3–2. Schematic diagram of Figure 3–1.

of the refrigerator light. This information cannot be determined from the schematic diagram (Figure 3–2). All that the schematic diagram shows is a common connection of the light, L1, and the fan to the black return power lead. It is not possible to determine from the schematic diagram that the connection is made at the light. Information of this type is only available from the pictorial diagram or wiring diagrams.

There is a trend by some manufacturers, particularly in the air-conditioning field, to combine pictorial and schematic diagrams. This combination has sometimes proven useful. The low-voltage transformer might be shown in a schematic diagram with the electrical symbols given as Figure 3–3*a*. The transformer might also be shown pictorially as in Figure 3–3*b*. Both the schematic diagram (Figure 3–3*a*) and the pictorial diagram (Figure 3–3*b*) present the information necessary in a diagram. With either it is readily understood that the device is a transformer and that the connections are made as coded. Further comparisons between the pictorial and schematic diagrams will be made later in this chapter.

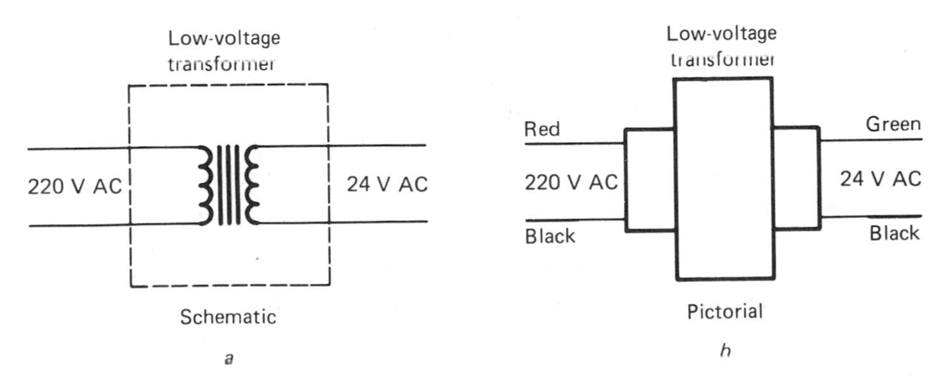

Figure 3–3. Schematic and pictorial diagrams of low-voltage transformers.

WIRING DIAGRAMS

The diagram that combines the best features of the pictorial and schematic diagrams is called a wiring diagram. The wiring diagram contains all the information that is to be given in diagrammatic form. It shows the components in the system in symbolic form, as well as the actual wiring connection, point to point. Figure 3–4 is a wiring diagram of the same refrigeration circuit shown in Figure 3–1 and Figure 3–2. In Figure 3–4, the identification of the components by symbol and the actual connection points are clear.

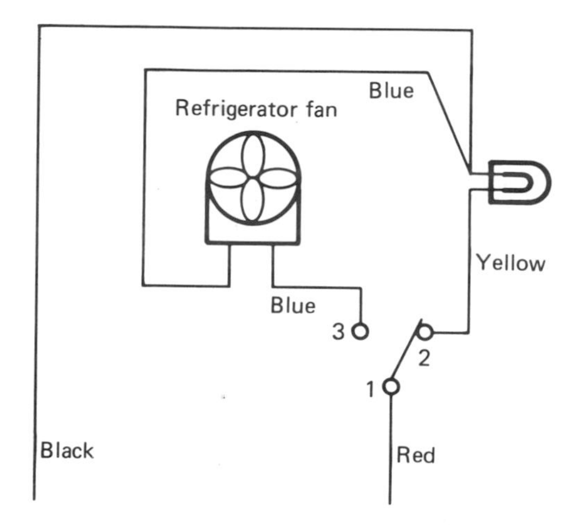

Figure 3–4. Wiring diagram of Figure 3–1.

The schematic and wiring diagrams are closely related, but the extra information, that is, the actual connection points of wires, is given only in the wiring diagram. It is readily seen from Figure 3–4 that the return connection from the fan is made at the light terminal, whereas all that can be determined from the schematic diagram (Figure 3–2) is that there is a common connection between the fan and the light.

The drawing of a wiring diagram of a larger system can become rather complex. In this situation, the wiring diagram may be furnished in sections. Figure 3–5 shows a section of a complete air-conditioning system. This wiring diagram must be furnished by the equipment manufacturer. The diagram includes only those components manufactured by the company. The remaining components will be furnished by the contractor completing the system. The wiring of the remainder of the system and the diagram of such wiring will be the responsibility of the contractor or architect designing the system.

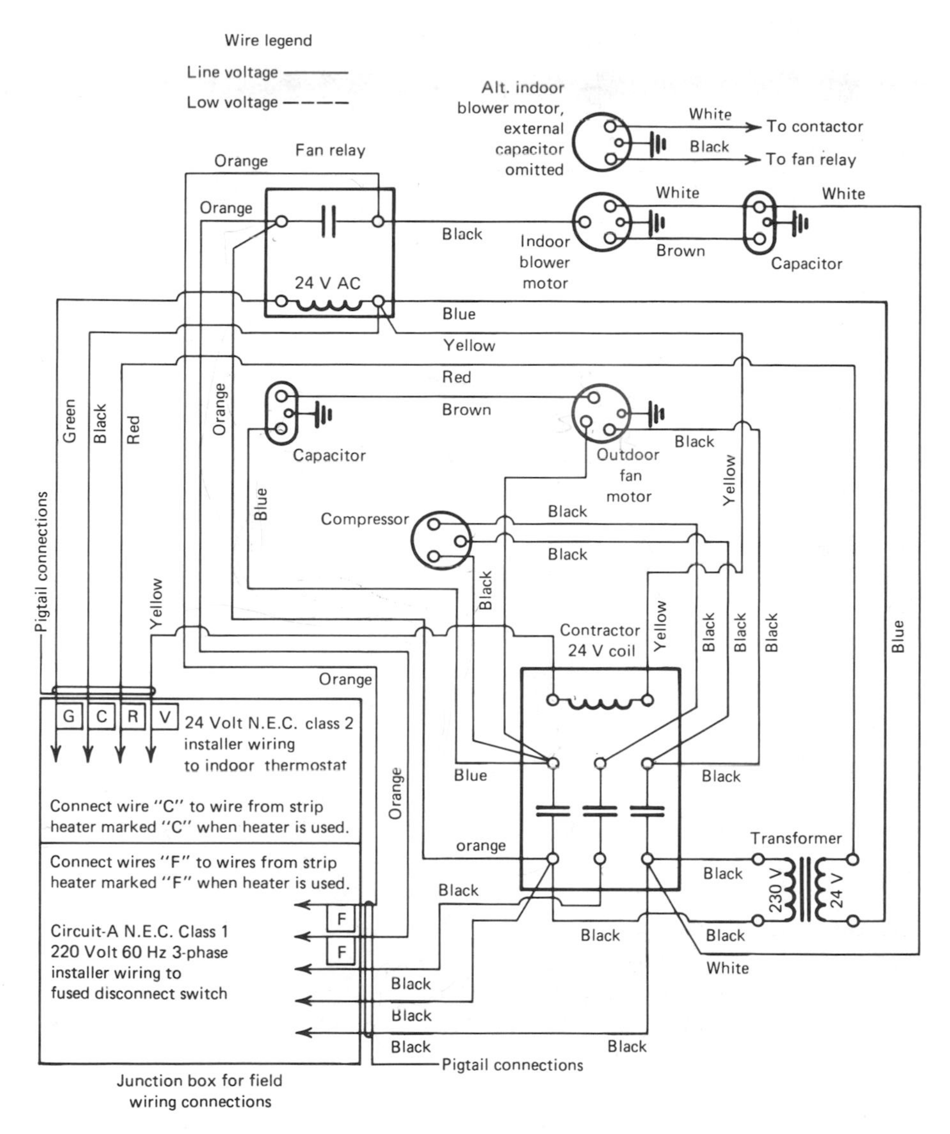

Figure 3–5. Wiring diagram of part of an air-conditioning system.

LADDER DIAGRAMS

The ladder diagram is a simplified schematic diagram that was developed for and is finding particular use in the air-conditioning trade area. Ladder diagrams present the connection of components in the most straightforward way. Such diagrams are therefore very useful in determining how a system

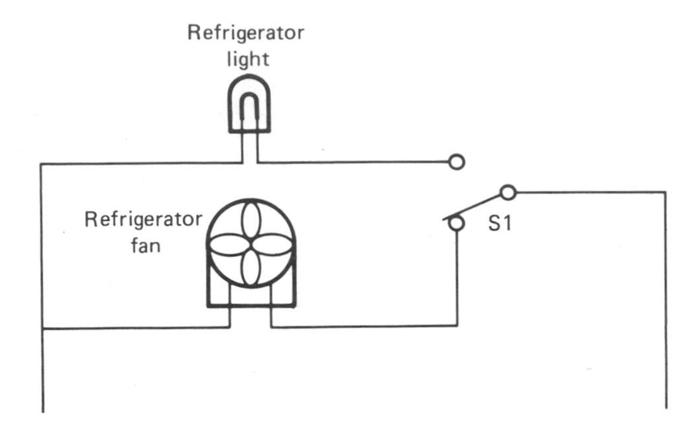

Figure 3–6. Ladder diagram of Figure 3–1.

operates. Figure 3–6 is a ladder diagram showing that section of the refrigerator shown in Figure 3–1 (the pictorial diagram), Figure 3–2 (the schematic diagram), and Figure 3–4 (the wiring diagram).

A ladder diagram presents the least amount of information in diagrammatic form but is the easiest for a technician to follow when trying to determine the sequence of operation of the system.

DIAGRAM COMPARISON

Figure 3–7 consists of a pictorial diagram (*a*) of a complete refrigerator and the schematic diagram (*b*) of the same unit. Pictorial diagrams and schematic diagrams are often supplied for refrigerators. When both diagrams are available, troubleshooting the system is a less difficult matter because more information is available to the technician.

When working with the air-conditioning system diagrams, the technician often encounters a combination of wiring diagrams and ladder diagrams. Usually, this combination provides the information necessary for installation and troubleshooting.

Compare the two parts of Figure 3–8. The system shown is a two-stage heat, two-stage cool air conditioner. In both the *a* (wiring diagram) and *b* (ladder diagram) sections of Figure 3–8, the thermostat is shown as a block. This is a common representation of a thermostat in wiring and ladder diagrams. A separate diagram of the thermostat is also provided (Figure 3–9).

If it is the purpose of the technician to determine how the system operates, then the ladder diagram is the best source of information. When it is necessary for the technician to follow a circuit from point to point, the wiring diagram is the diagram to use.

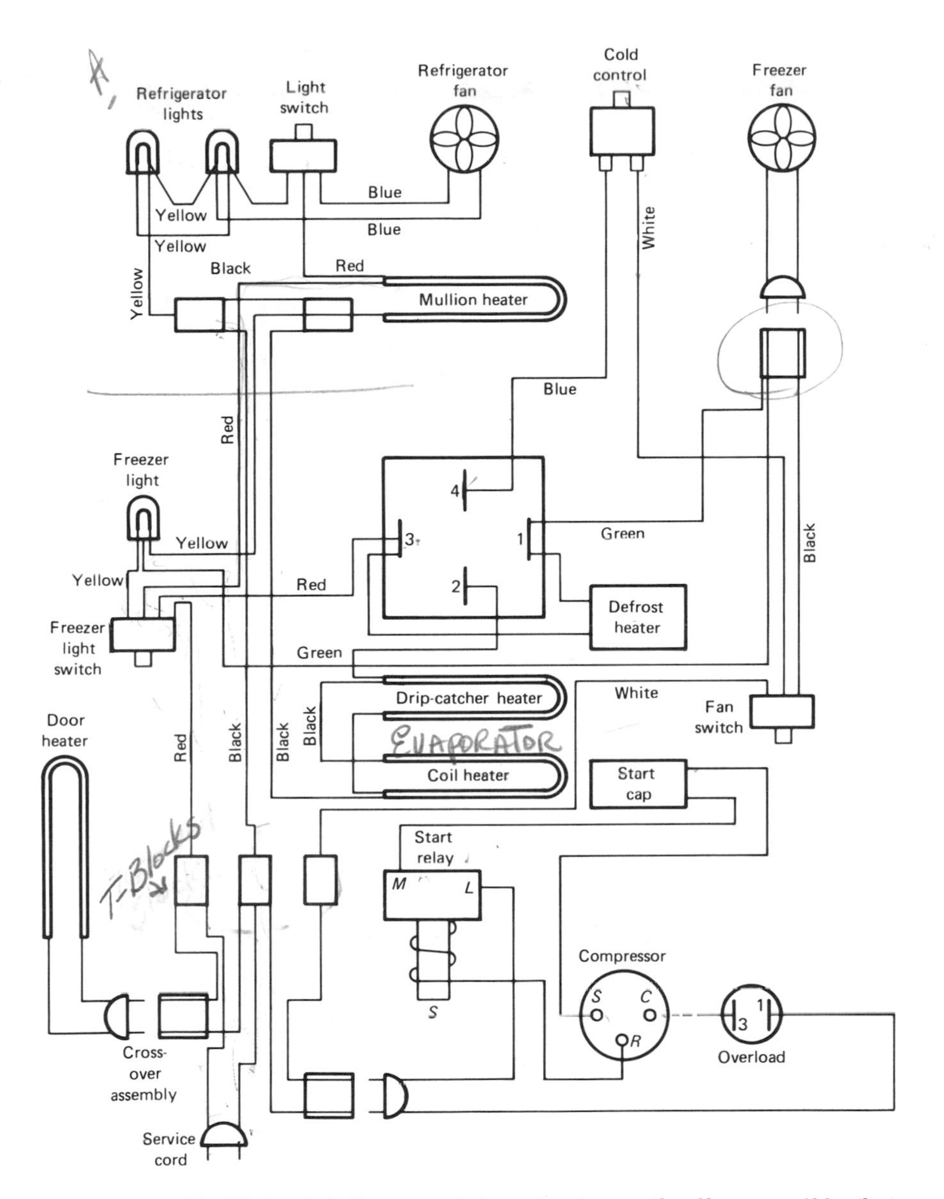

Figure 3–7. Pictorial diagram (*a*) and schematic diagram (*b*) of refrigerator.

Although it may be necessary for the technician to refer to a ladder diagram while troublehsooting, the actual connection points for test instruments must be determined using the wiring diagram.

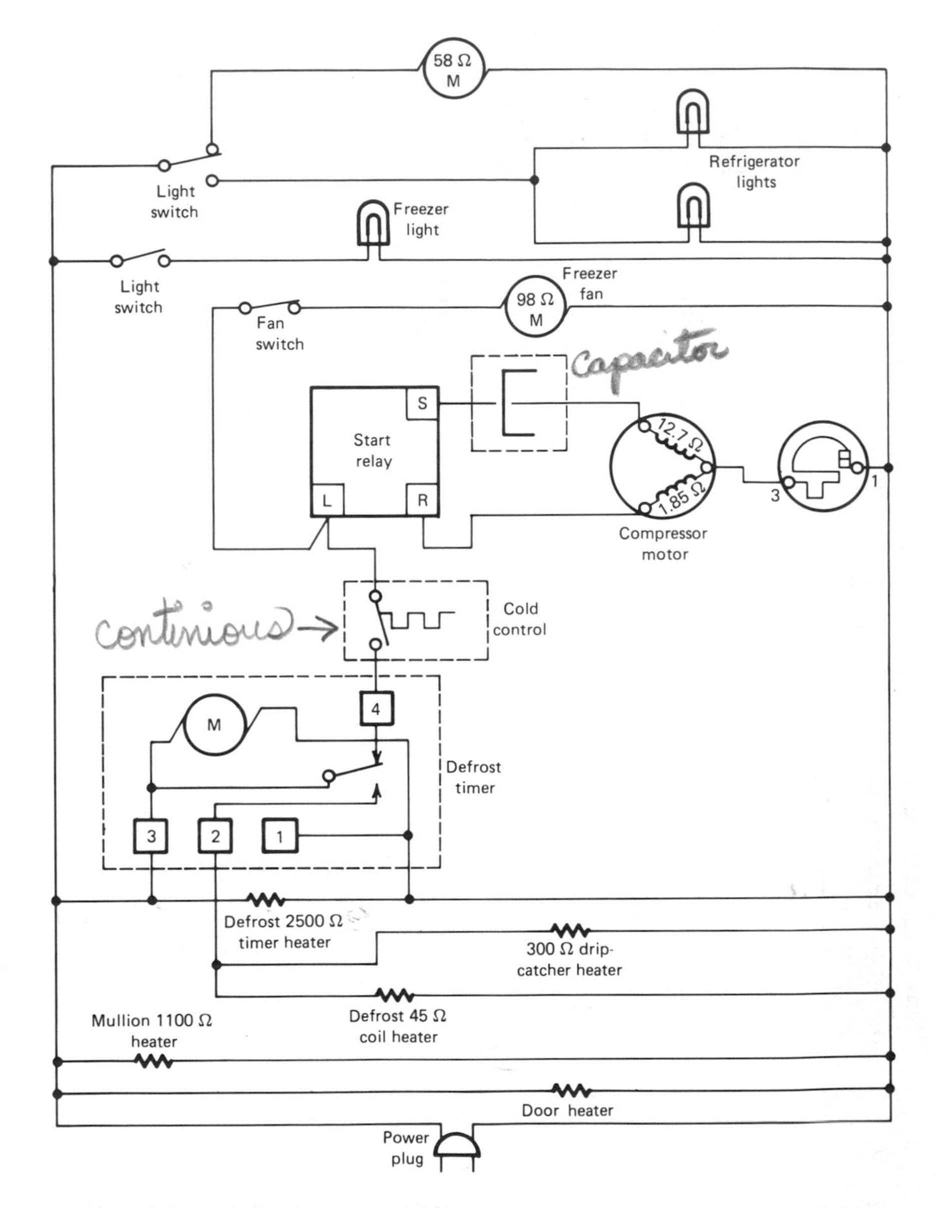
58 Ω
M
Refrigerator lights
Light switch
Freezer light
Light switch
Freezer fan
98 Ω
M
Fan switch
Capacitor
S
Start relay
L
R
12.7 Ω
1.85 Ω
Compressor motor
3
1
continious
Cold control
4
M
Defrost timer
3
2
1
Defrost 2500 Ω timer heater
300 Ω drip-catcher heater
Defrost 45 Ω coil heater
Mullion 1100 Ω heater
Door heater
Power plug
Schematic

Figure 3–7(b).

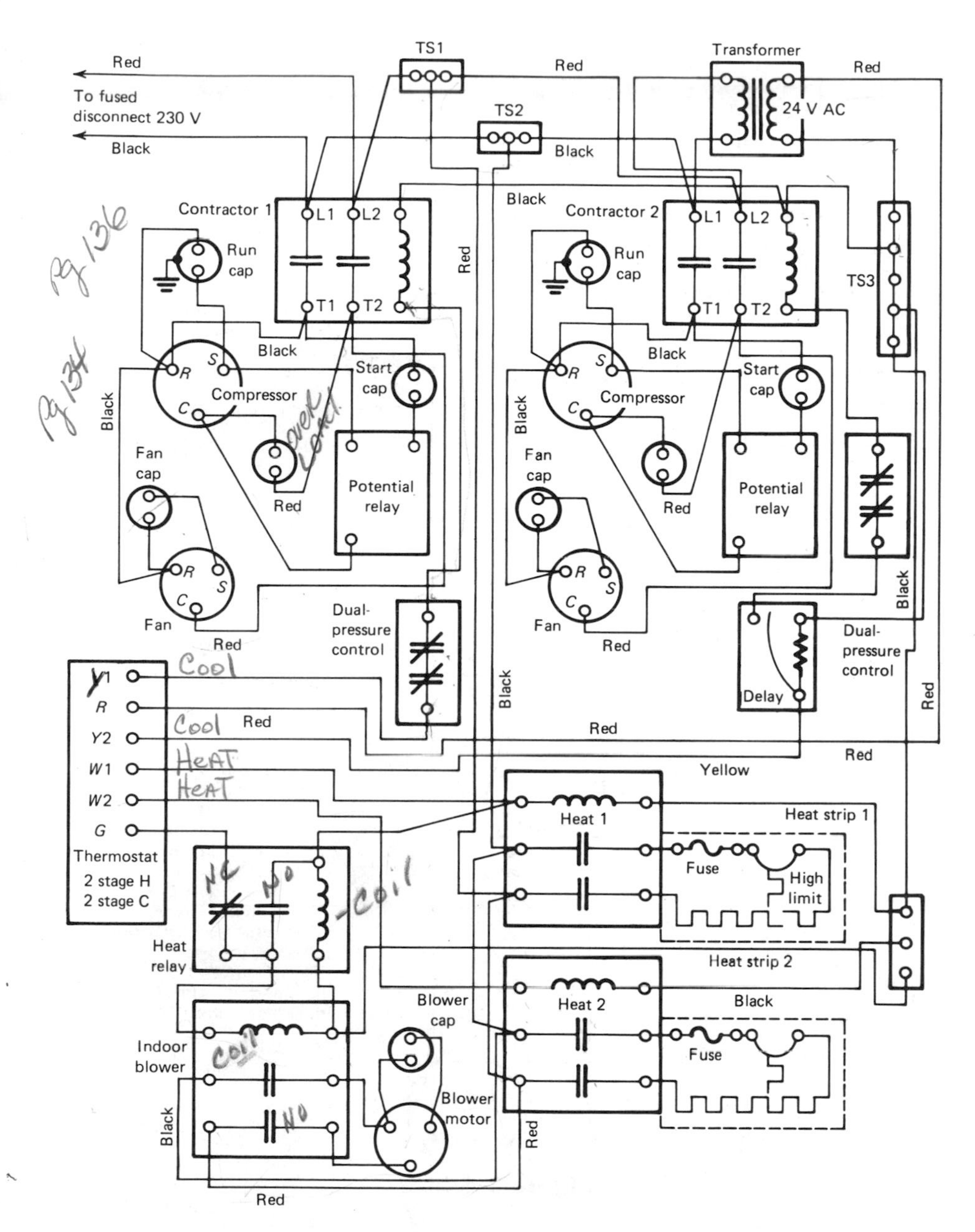

Figure 3–8. Two-stage heat, two-stage cool air conditioner.

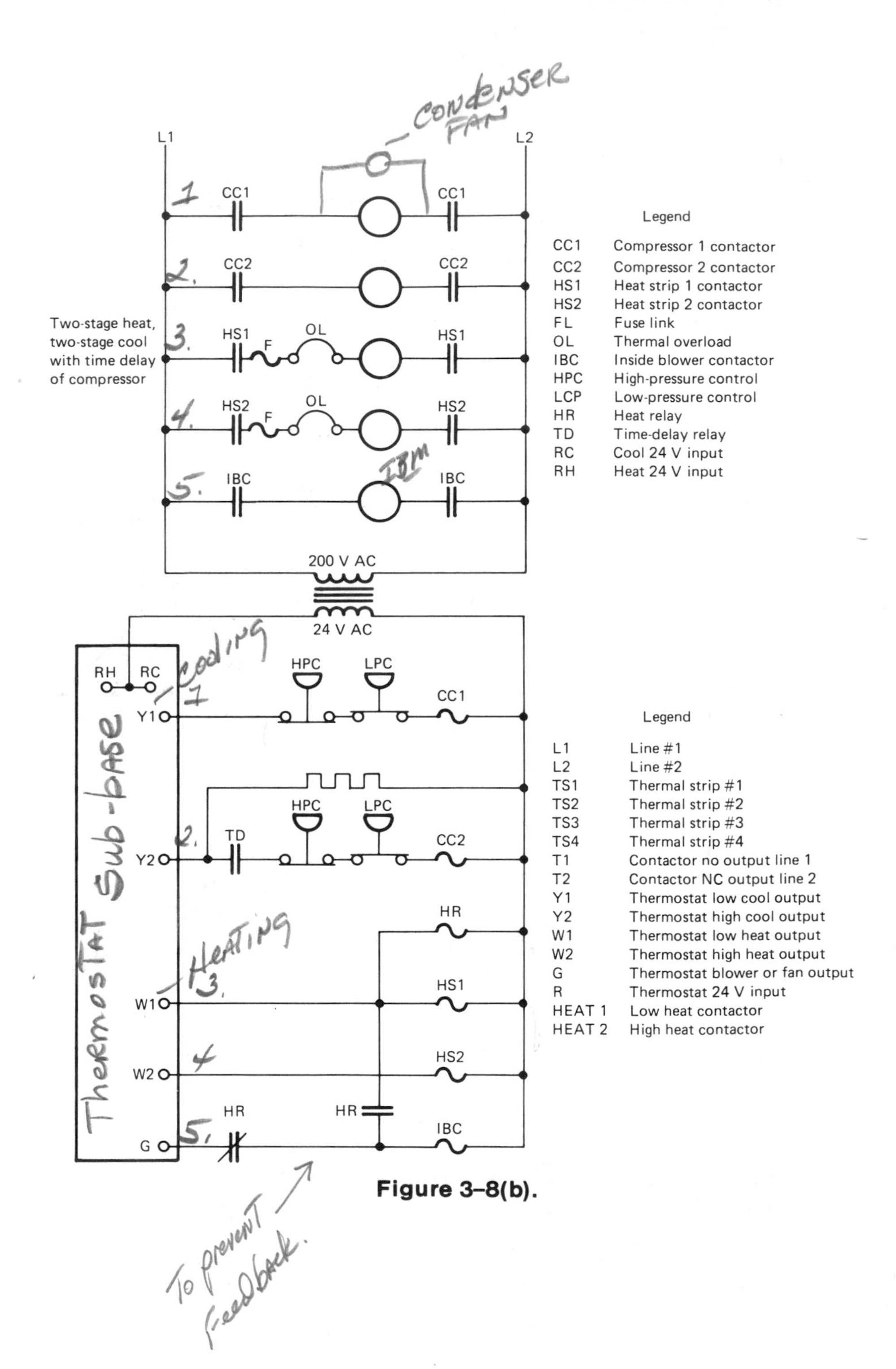
L1
L2
CONDENSER FAN
1
CC1
CC1
2.
CC2
CC2
Two-stage heat, two-stage cool with time delay of compressor
3.
HS1
F
OL
HS1
4.
HS2
F
OL
HS2
5.
IBC
IBM
IBC
200 V AC
24 V AC
Legend
CC1 Compressor 1 contactor
CC2 Compressor 2 contactor
HS1 Heat strip 1 contactor
HS2 Heat strip 2 contactor
FL Fuse link
OL Thermal overload
IBC Inside blower contactor
HPC High-pressure control
LCP Low-pressure control
HR Heat relay
TD Time-delay relay
RC Cool 24 V input
RH Heat 24 V input
Thermostat Sub-base
RH
RC
Cooling
1
Y1
HPC
LPC
CC1
2.
Y2
TD
HPC
LPC
CC2
HR
Heating
3.
W1
HS1
4
W2
HS2
5.
G
HR
HR
IBC
To prevent feedback.
Legend
L1 Line #1
L2 Line #2
TS1 Thermal strip #1
TS2 Thermal strip #2
TS3 Thermal strip #3
TS4 Thermal strip #4
T1 Contactor no output line 1
T2 Contactor NC output line 2
Y1 Thermostat low cool output
Y2 Thermostat high cool output
W1 Thermostat low heat output
W2 Thermostat high heat output
G Thermostat blower or fan output
R Thermostat 24 V input
HEAT 1 Low heat contactor
HEAT 2 High heat contactor

Figure 3–8(b).

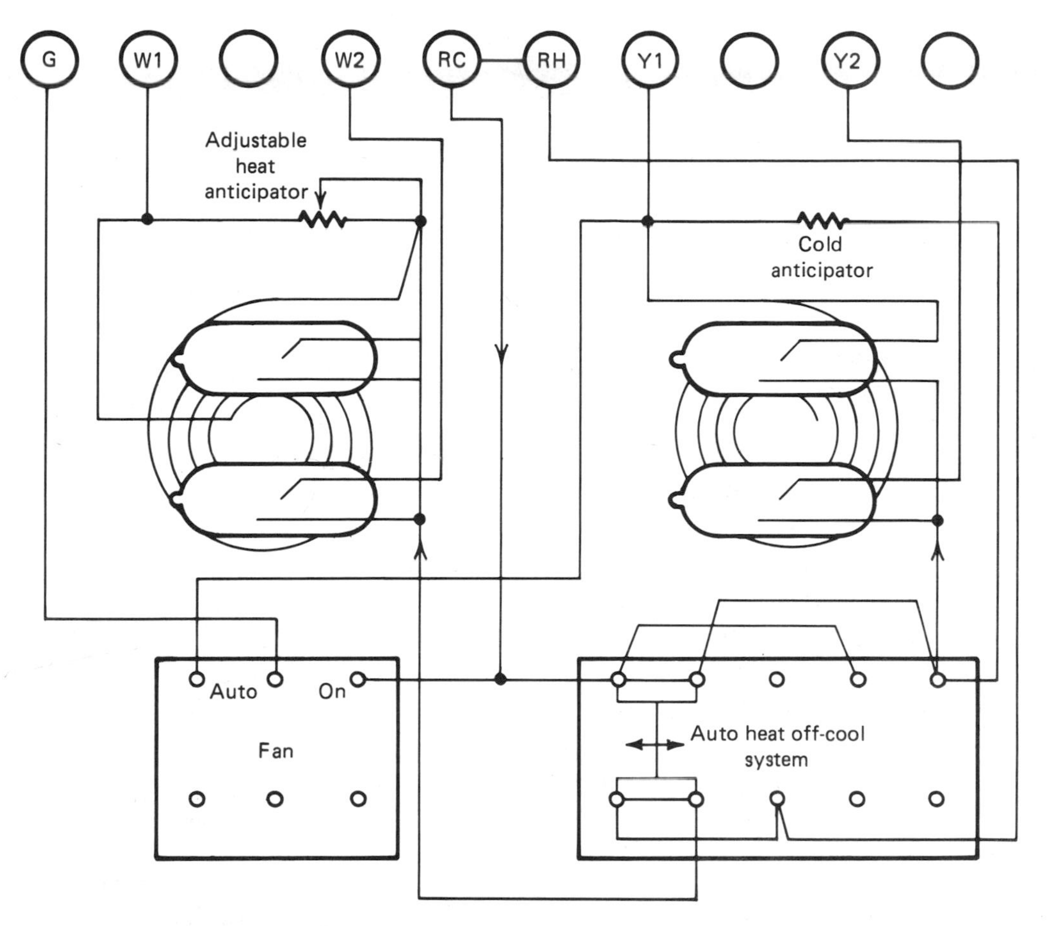

Figure 3–9. Thermostat of two-stage heat, two stage cool air conditioner.

Review Questions

Chapter 3

Choose from the following words to fill in the blanks in these questions:

(a) Pictorial
(b) Schematic
(c) Wiring
(d) Ladder

1. The diagram that is most useful in determining system operation is the ___ladder___ diagram.
2. The diagram that includes symbols of the components and the indication of actual connection points is the ___Pictorial___ diagram.
3. The diagram that includes pictures of the components, an indication of color-coded wires, and wire routes is the ___wiring___ diagram.
4. The diagram that uses symbols of the components and shows the electric circuit but not the connection points is called the ___schematic___ diagram.
5. Two diagram types are most often used with air-conditioning systems. They are the ___Wiring___ and the ___ladder___ diagrams.
6. Two diagram types are most often used with refrigeration circuits. They are the ___Pictorial___ and the ___Schematic___ diagrams.

Chapter 4

THE LADDER DIAGRAM

As indicated in Chapter 3, the ladder diagram is the easiest to follow because it presents the connection of the major electrical components in a straightforward manner. Only the connections of components that operate or control are shown.

In this chapter, a ladder diagram of an air-conditioning system and a refrigerator will be developed, beginning with the electrical connection of a motor. Controls switches and relays will be added until a complete air-conditioning system is diagrammed. As the components are added, they will be explained as to purpose and function.

THE HIGH-VOLTAGE SYSTEM

Figure 4–1 is a ladder diagram of the connection of an electric motor to a power source. There are no switches in the circuit. The only control of motor operation is the application of electrical power. If the required voltage is present between L1 and L2, the motor will run. If the required voltage is not present between L1 and L2, the motor will not run.

In Figure 4–2, a simple switch is shown added in series with the L1 line to the motor. Two conditions must now be met before the motor will operate. First, the required voltage must be available between L1 and L2; second, the

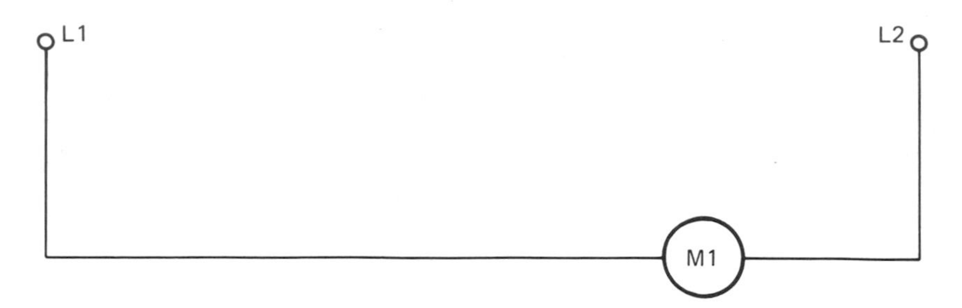

Figure 4–1. Electric motor connected to power source.

Figure 4–2. Addition of switch in series.

switch in series with L1 must be closed. If either condition is not met, the motor will not run.

If a second motor, M2, is added in parallel with the first motor, M1, both motors will operate together. If the required voltage is present between L1 and L2 and the switch is closed, both motors will operate. These conditions are shown in Figure 4–3. If either the line voltage is removed or the

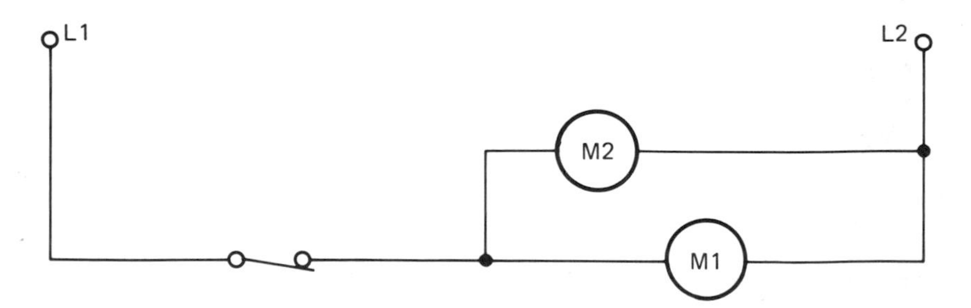

Figure 4–3. Addition of second motor in parallel.

switch is opened, the motors will not operate. These same conditions exist for a third motor, M3, connected in parallel with M1 and M2, as shown in Figure 4–4.

In an air-conditioning system, the motors have a purpose and therefore

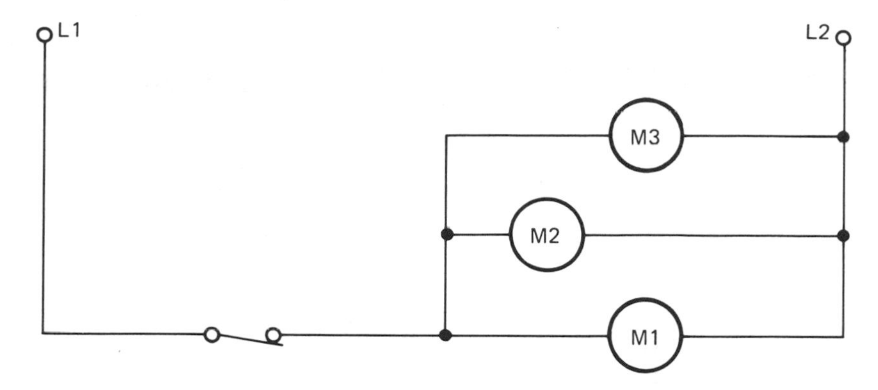

Figure 4–4. Addition of third motor in parallel.

a name. In this case, the first motor will be designated as the outside fan motor and will be coded OFM. The second motor is designated as the compressor motor and is coded CM. The third motor is the inside blower motor, coded IBM. The switch is a set of contacts in the control thermostat, coded CT (see Figure 4–5).

The circuit is now that of a simple air-conditioning system where all motors are controlled by the set of contacts in the thermostat. The condition of the thermostat contacts (open or closed) is, of course, a function of the temperature at the location of the thermostat. If the temperature is above the required temperature, the thermostat contacts (CT) will be closed. The three motors (IBM, CM, and OFM) will operate as shown in Figure 4–5.

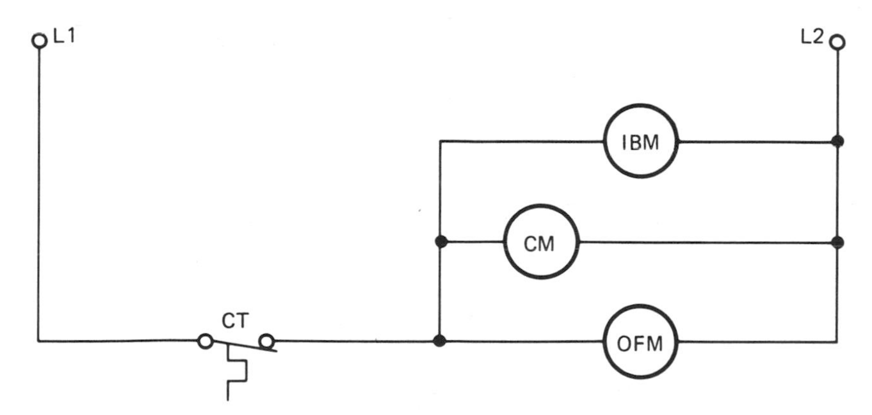

Figure 4–5. Operation of circuit with closed thermostat contacts.

When the temperature at the control thermostat becomes lower than the thermostat setting, the thermostat contacts (CT) will open. The circuit is shown in Figure 4–6. It is obvious that the three motors (IBM, CM, and OFM) will stop when the control thermostat opens, thus removing power

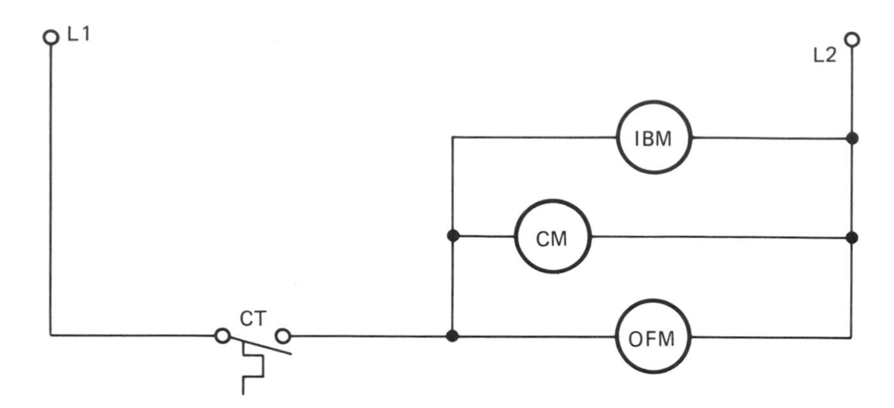

Figure 4–6. Operation of circuit with open thermostat contacts.

from the motor terminals. The motors will remain off until the room temperature rises, causing the control thermostat contacts to close. Power will again be applied to the motors, causing them to rotate.

The purpose of the high-pressure control (HPC) was given in Chapter 2. An application of this control is shown in Figure 4–7. The contacts of the HPC are normally closed and are connected in series with the line feeding

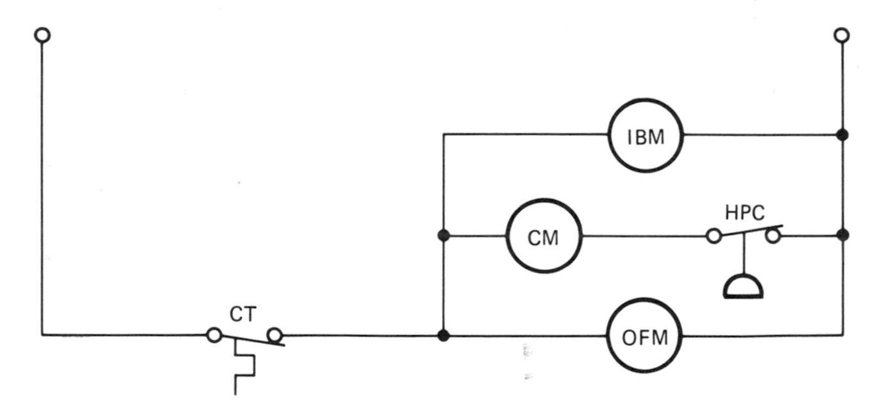

Figure 4–7. Operation of high-pressure control.

the compressor motor. If, for any reason, the pressure in the high side of the compressor exceeds the preset limit, the contacts of the HPC will open, removing electrical power from the compressor motor.

LOW-VOLTAGE CONTROL OF THE HIGH-VOLTAGE SYSTEM

The devices that have been shown controlling the operation of the air-conditioning system are all in a high-voltage system. In actual practice, most control devices operate at low voltage, usually 24 volts. The low-voltage system operates relays and contactors that control the operation of high-voltage devices.

In the following paragraphs, the ladder diagram of such a system will be developed. Only the active components will be shown. Items such as fuses and overloads will not be included.

The system to be developed is a two-stage heat, two-stage cool air-conditioning system. The operation of the thermostat used with this system is decribed in terms of the symbols given in Chapter 2. It is important that the operation of the thermostat be understood before the reader attempts to follow the ladder diagram of a complex air-conditioning system such as the one being presented.

TWENTY-FOUR-VOLT SIGNAL POWER

In an air-conditioning system, the 24 volts AC signal power is obtained from a step-down transformer. The transformer's primary is connected between L1 and L2 of the high-voltage source. The low voltage (24 volts AC) is obtained from the secondary. The 24 volts AC will be available across the secondary as long as the high voltage is present between L1 and L2 and therefore across the primary. This circuit is shown in Figure 4–8.

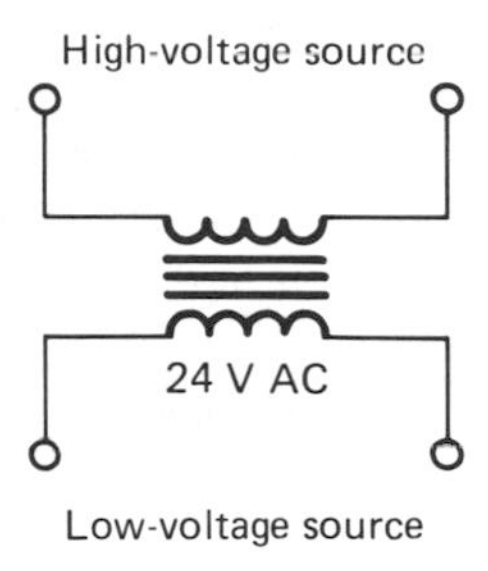

Figure 4–8. Control transformer.

HIGH-VOLTAGE COMPONENTS AND CONTACTORS

The purpose of the control system is to ensure proper application of high voltage to the operating elements of the system. Before the ladder diagram can be attempted, a complete list of operating components must be made. Each of the operating components will be controlled by a contactor circuit. The contactor coils will be controlled by the low-voltage circuit.

In a two-stage heat, two-stage cool air-conditioning system, the components to be controlled are:

1. Heat strip 1
2. Heat strip 2
3. Compressor 1
4. Compressor 2
5. Condenser fan 1
6. Condenser fan 2
7. Inside blower motor

The components listed are all operated directly across the high-voltage lines, L1 and L2. Figure 4–9 shows the connection system for these components and gives the component name and contractor name. Both the components to be controlled and the primary of the low-voltage transformer are connected directly across the high-voltage lines.

The operation of the components in the high-voltage circuit is easy to understand. Whenever the contactor associated with a component is energized, the contactor will be closed, and high-voltage power will be applied to the component. If the contactor is not energized, power will not be applied to the component. Essentially, this means that

1. If compressor 1 contactor (CC1) is energized, its contacts will be closed and compressor 1 will run.
2. If compressor 2 contactor (CC2) is energized, its contacts will be closed, and compressor 2 will run.
3. If the heat strip 1 contactor (HS1) is energized, its contacts will be closed. Power will be applied to the heat strip, and electrical energy will be converted to heat energy.
4. If the heat strip 2 contactor (HS2) is energized, its contacts will be closed. Power will be applied to the heat strip, and it will heat up.
5. If the inside blower contactor is energized, its contacts will be closed. Power will be applied to the inside blower motor, and it will rotate.
6. Power will be applied to the primary of the low-voltage transformer as long as power is available between L1 and L2.

It is obvious from the last few paragraphs and from the ladder diagram of the high-voltage circuits (Figure 4–9) that with the exception of the low-voltage transformer, which is always connected across the line, the contactors control the application of high-voltage power to the operating motors and heat strips. The motors and heater elements are connected in the most straightforward fashion. The circuits are easy to follow, and there should not be a problem in determining whether power is applied to a component. If the contactor is energized, power will be available at the component terminals. If the contactor is de-energized, power will not be available at the component terminals.

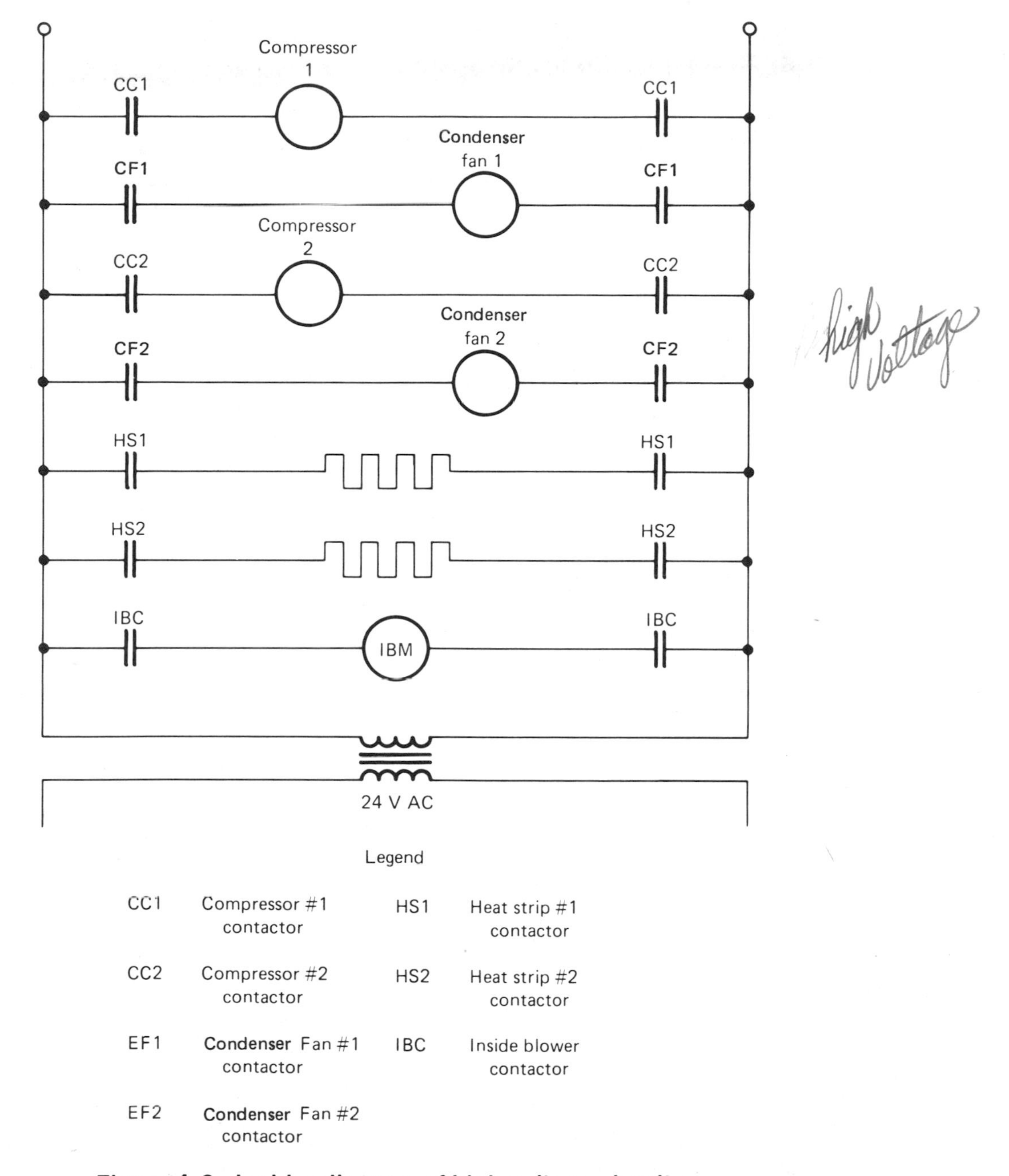

Figure 4–9. Ladder diagram of high-voltage circuits.

LOW-VOLTAGE CIRCUIT CONTROL OF THE HIGH-VOLTAGE CONTACTOR

Here in the low-voltage circuit is where actual control of the air-conditioner takes place. The switches in the thermostat, pressure switches that are connected to the compressor system, time-delay circuits, and relay contacts are interconnected in the low-voltage system. These switches and contacts determine whether 24 volts AC will be applied to the coils of the contactors that control the application of high-voltage power to the compressor motors, heat strips, and inside blower motors.

COMPLETING THE LADDER: LOW VOLTAGE

As shown in Figure 4–9, the primary of the low-voltage transformer is connected across the high-voltage lines. Twenty-four volts AC is available across the secondary whenever power is connected to the air-conditioning system.

THERMOSTATS

The thermostat is shown in the ladder diagram as a rectangle with input and output terminals. The 24 volt control current is fed from the transformer low-voltage secondary to the input of the thermostat on the red heat (RH) and red cool (RC) terminals. This is shown in Figure 4–10. The 24 volt outputs of

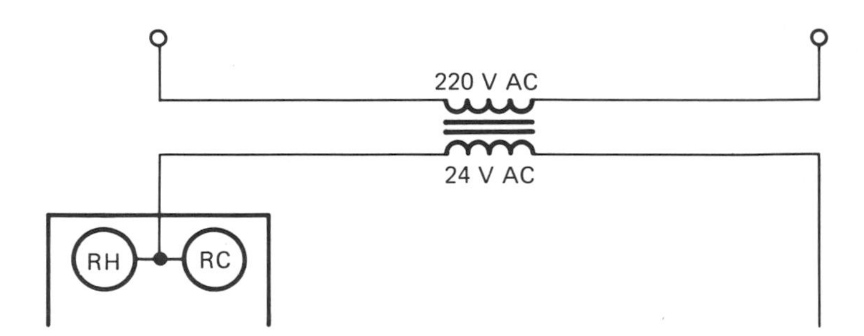

Figure 4–10. Input of 24 volts AC to thermostat.

the thermostat are at the Y1, Y2, W1, W2, and G terminals (see Figure 4-11). The following should be considered with reference to the ladder diagram:

a. Twenty-four volts AC will be available at terminal *Y1* if "cool" is selected and operation of a single compressor is called for. The wire connected to *Y1* is usually yellow.

b. Twenty-four volts AC will be available at terminals *Y*1 and *Y* 2 if "cool" is selected and operation of dual compressors is called for. The wired connected to *Y*2 is usually yellow striped.

c. Twenty-four volts AC will be available at terminal *W*1 if "heat" is selected and operation of a single heat strip is called for. The wire connected to *W*1 is usually white.

d. Twenty-four volts AC will be available at terminals *W*1 and *W*2 if "heat" is selected and operation of dual heat strips is called for. The wire connected to *W*2 is usually white striped.

e. Twenty-four volts AC will be available at *G* if the fan switch of the thermostat is in the "on" position or when "cool" is selected and the operation of a compressor is called for. The wire connected to *G* is usually green.

THERMOSTAT CONNECTIONS IN THE LADDER DIAGRAM

In Figure 4–11, the thermostat is shown with output terminals *Y*1, *Y*2, *W*1, *W*2, and *G*. The next step in completion of the low-voltage ladder diagram is to add the connections to the *G* output for inside fan or blower motor control.

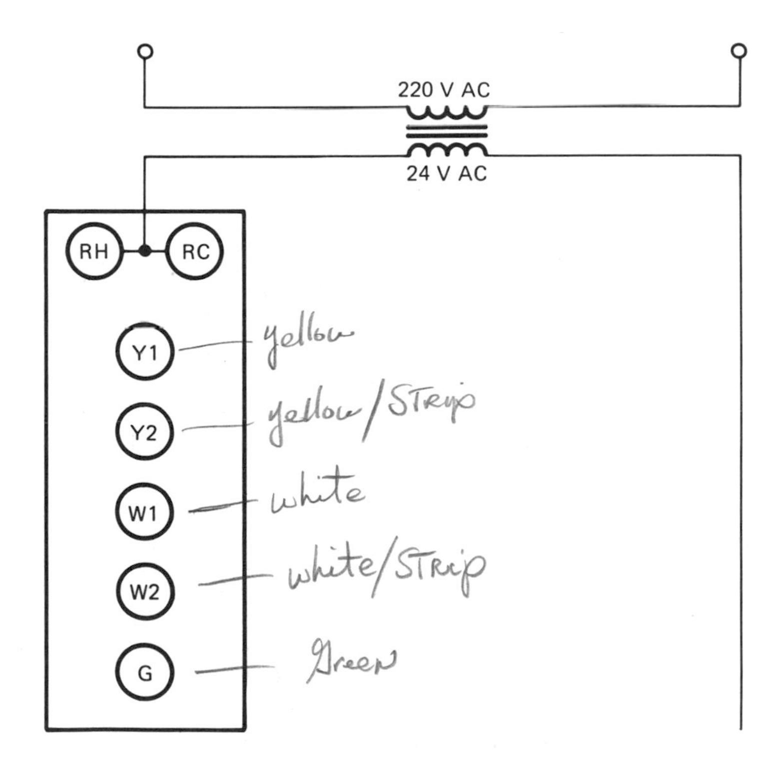

Figure 4–11. Input of 24 volts AC to thermostat with thermostat outputs.

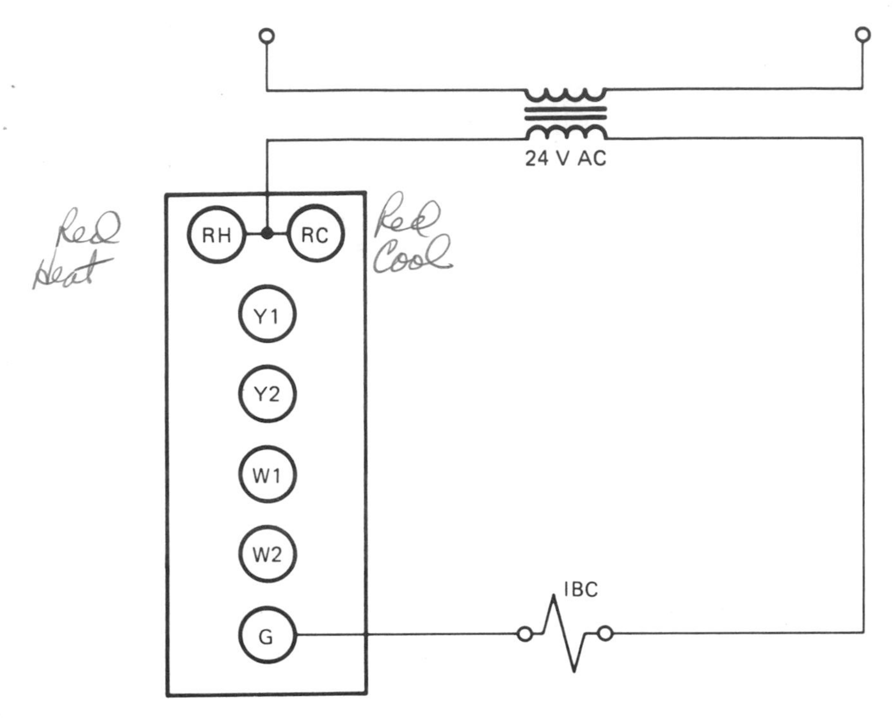

Figure 4–12. Connection of inside blower contactor coil.

FAN OR BLOWER MOTOR CONTROL

Figure 4–12 shows the connection of the coil of the fan contactor, designated as the inside blower contactor (IBC). As already indicated, 24 volts AC is available at the *G* terminal if the fan switch of the thermostat is in the "on" position, or if "cool" is selected and the operation of a compressor is called for. Whenever 24 volts AC is available at the *G* terminal of the thermostat, the IBC will be energized. The contacts of the inside blower contactor will close, and high-voltage power will thus be applied to the inside blower motor. (The high-voltage circuit is shown in Figure 4–9.)

COMPRESSOR 1

The control of compressor 1 requires consideration of the temperature. The thermostat provides an output at Y1 when the temperature is above the selected temperature. The 24 volts AC at the Y1 output will cause the compressor 1 contact coil to energize. The circuit is shown in Figure 4–13. Whenever compressor 1 is operating, condenser fan 1 should also operate. The contactor coil is shown connected in Figure 4–14. Dual compressor air-conditioning systems are of a horsepower rating that requires the use of

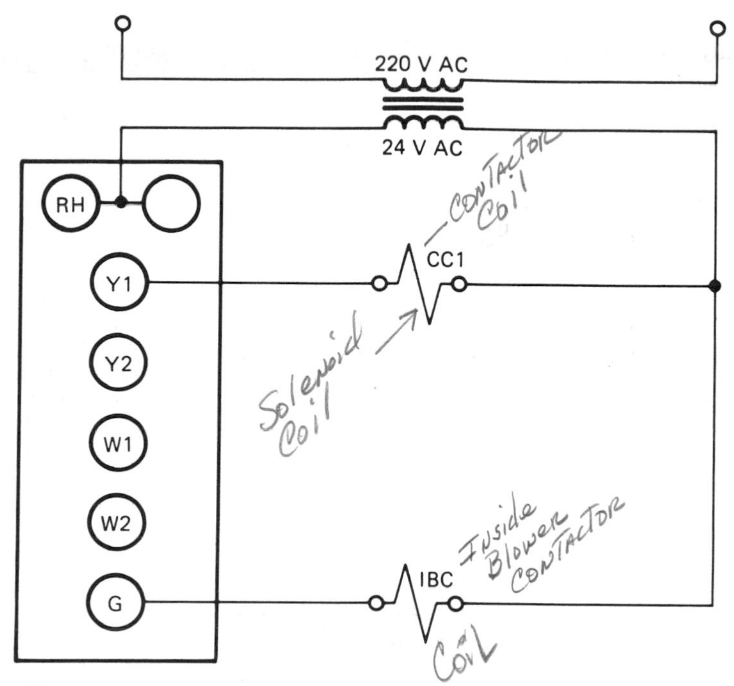

Figure 4–13. Addition of compressor 1 contactor coil.

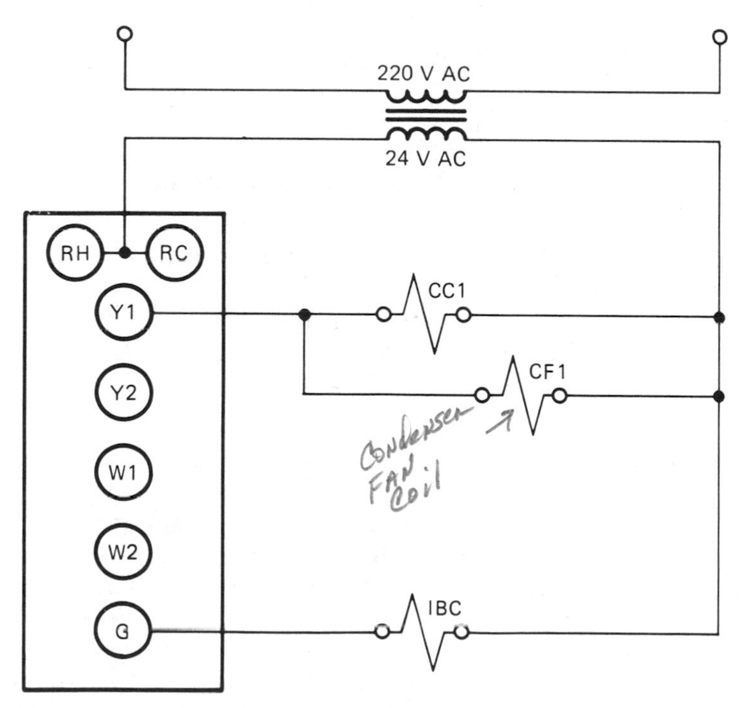

Figure 4–14. Addition of condenser fan 1 contactor coil.

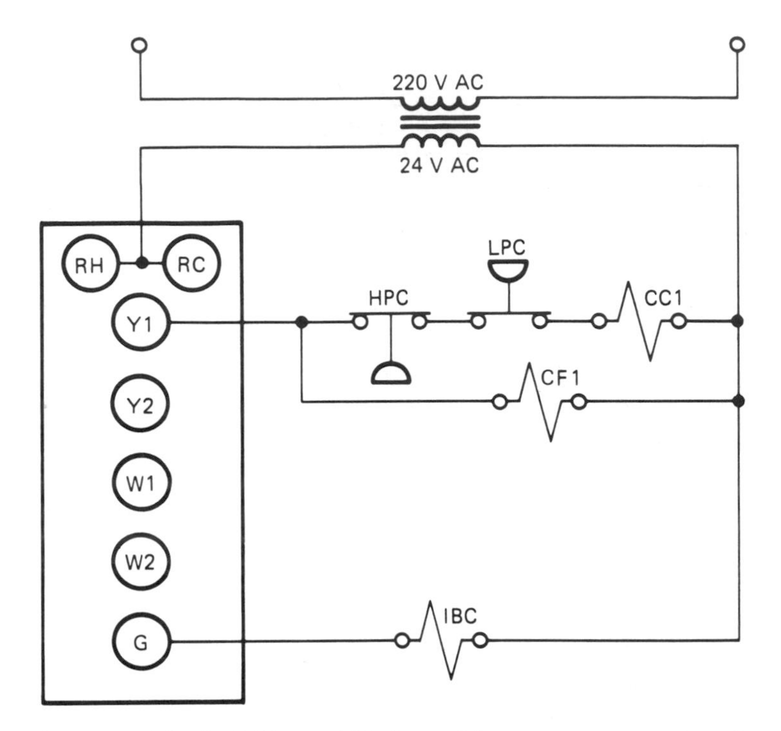

Figure 4–15. Addition of high- and low-pressure cotnrols.

pressure controls—usually both a high-pressure control (HPC) and a low-pressure control (LPC). In Figure 4–15, both the HPC and the LPC are shown connected in series with the coil of CC 1. The HPC and the LPC do not affect the connection to condenser fan 1. They do provide control of the compressor 1 contactor coil (CC1). Should the high-side pressure increase above the preset limit, the HPC will open, removing 24 volts AC from the compressor 1 contactor coil. Similarly, if the low-side pressure should be reduced below the preset limit, the LPC will open, removing 24 volts AC from the compressor 1 contactor coil. (Both pressure controls are shown in the open position in Figure 4–16.)

COMPRESSOR 2

Compressor 2 is controlled by the compressor 2 contactor (CC2). The controls regulating the operation of the compressor 2 contactor coil are the same as those controlling the compressor 1 contactor coil: the HPC and the LPC, with the addition of the time-delay contacts. The controls are shown in Figure 4–17. The HPC and the LPC provide the same protection in compressor 2 circuits as they do in compressor 1 circuits. The time-delay contact

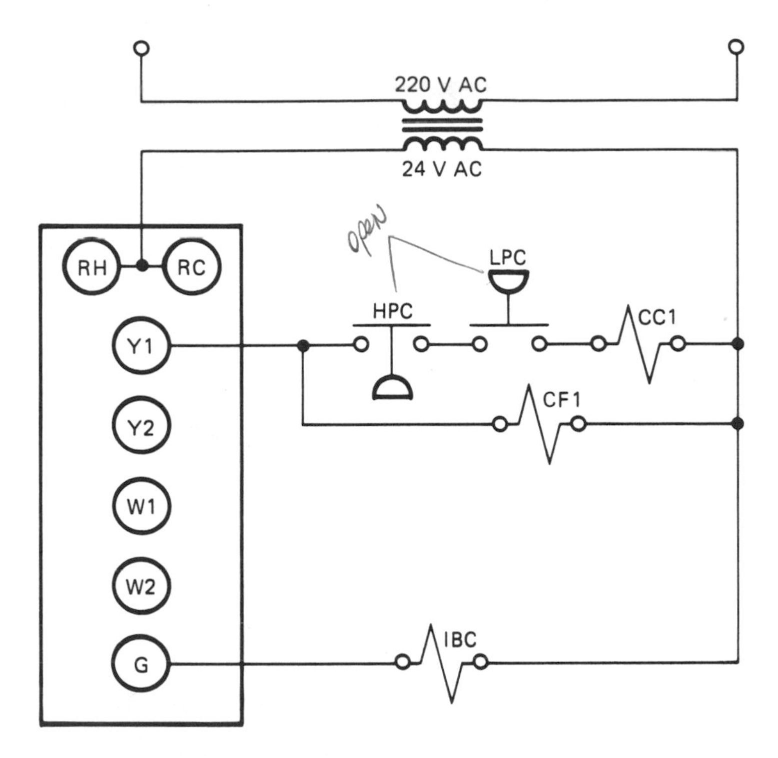

Figure 4–16. Effect of opening of high- and low-pressure controls on contactor coil circuit.

ensures that both the compressor 1 contactor and the compressor 2 contactor do not close at the same time. If both motors were to start at the same time, the line would have to provide the high starting current of both motors at the same time. If the starting of the two compressors is staggered, the high current is limited to the starting current of one motor.

As shown in Figure 4–17, the contacts of the time-delay relay are connected in series with the HPC and the LPC. The heater of the time-delay relay is connected from the *Y*2 terminal to the 24 volts AC common. If the thermostat setting calls for maximum cool, with both compressors on, and 24 volts AC becomes available at *Y*1 and *Y*2 of the thermostat at the same time, compressor 1 will energize immediately.

The operation of the compressor 2 contactor will be delayed until the 24 volts AC available at *Y*2 of the thermostat heats the time-delay relay heater. After a preset period, the contacts of the time-delay relay will close. The 24 volts AC from *Y*2 will be fed through the time-delay contacts and through the high- and low-pressure contacts to the compressor 2 coil. The compressor 2 contactor will close a short time after the compressor 1 contactor. The compressor 2 condenser fan contactor is connected after the

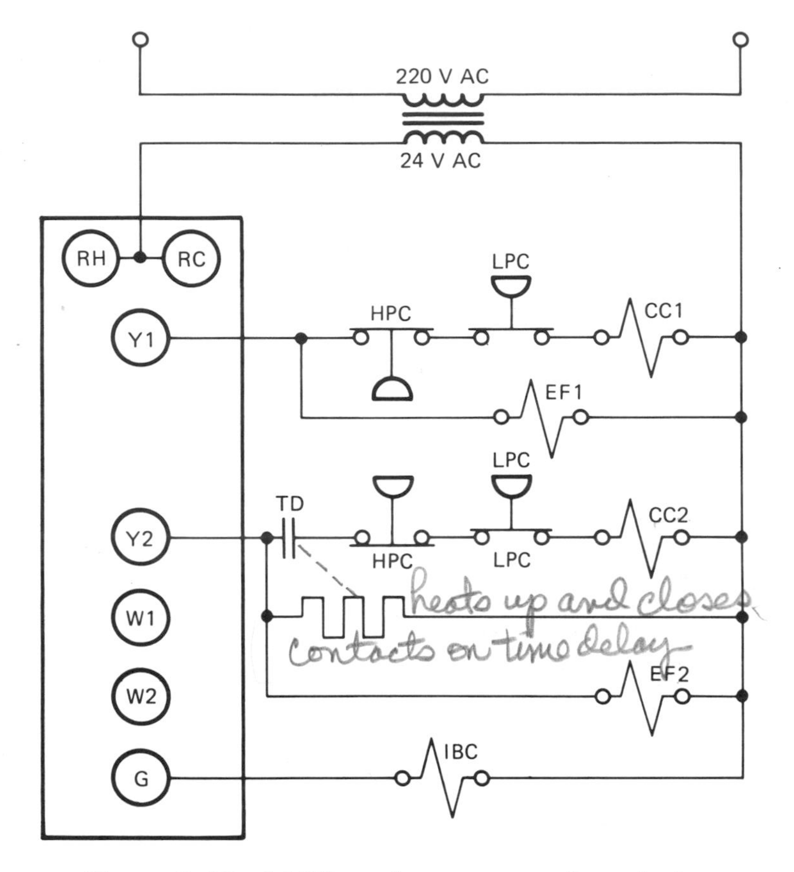

Figure 4–17. Addition of compressor 2 controls.

time-delay contacts close. The compressor 2 evaporator fan will come on at the same time as compressor 2.

HEAT STRIP CONTACTOR CONTROL

The operation of the heat contactors is straightforward. When 24 volts AC is available at *W*1, it will be applied to the coil of the heat strip 1 contactor, and the contactor will energize. When 24 volts AC is available at *W*2, it will be applied to the heat strip 2 contactor coil and the contacts will energize.

The circuitry requiring investigation is that of the inside blower motor (IBM). The inside blower motor must operate whenever a heat strip is operating. But if the contactor coil of the inside blower motor is connected directly to *W*1 as shown in Figure 4–18, two problems will result:

1. During cooling operations, whenever the inside blower motor is operating, heat strip 1 will be on.

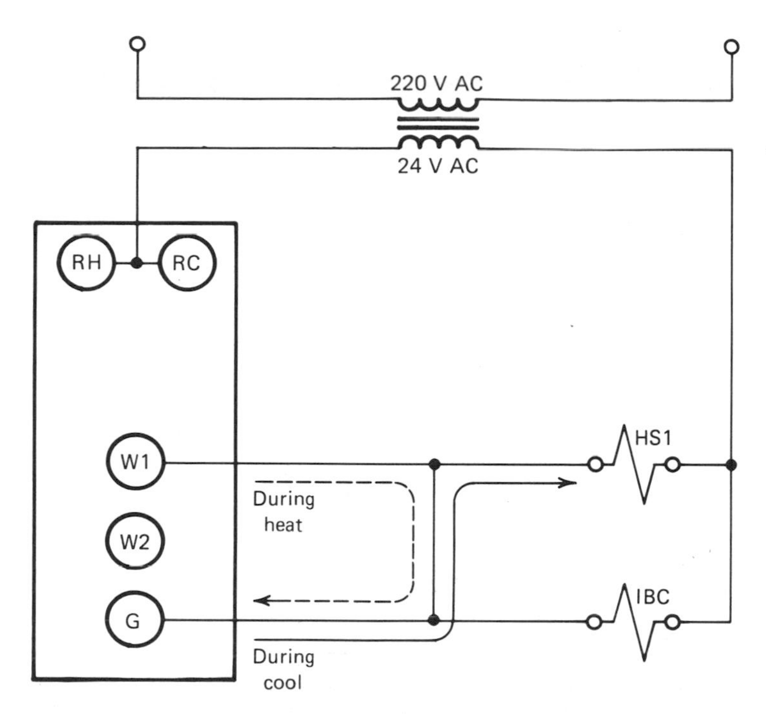

Figure 4–18. Possible sneak circuit if IBC coil and heat contactor coil are wired together.

2. During heating operations, there is a possible sneak circuit back through the thermostat that might turn on compressor 1.

In order to eliminate these problems, a heat relay is added in the low-voltage circuit to control the inside blower motor during heating operations. This is shown in Figure 4–19.

The normally closed contacts of the heat relay provide a connection from the *G* output of the thermostat to the inside blower contactor during cooling operations. When "heat" is selected, 24 volts AC is available at the *W*1 terminal of the thermostat. This 24 volts AC is fed to the coil of the heat strip 1 contactor as shown in Figure 4–18 and is also fed to the heat relay coil as shown in Figure 4–19. Whenever "heat" is selected and required, the heat relay will be energized, providing the necessary control of the inside blower motor through control of the inside blower contactor coil.

As shown in Figure 4–19, the normally open contacts of the heat relay close, providing a complete circuit to the IBC coil. The contactor energizes, and the inside blower motor rotates. At the same time, the normally closed contacts of the heat relay open. This eliminates the possibility of a sneak circuit back through the thermostat to the compressor contactor coils.

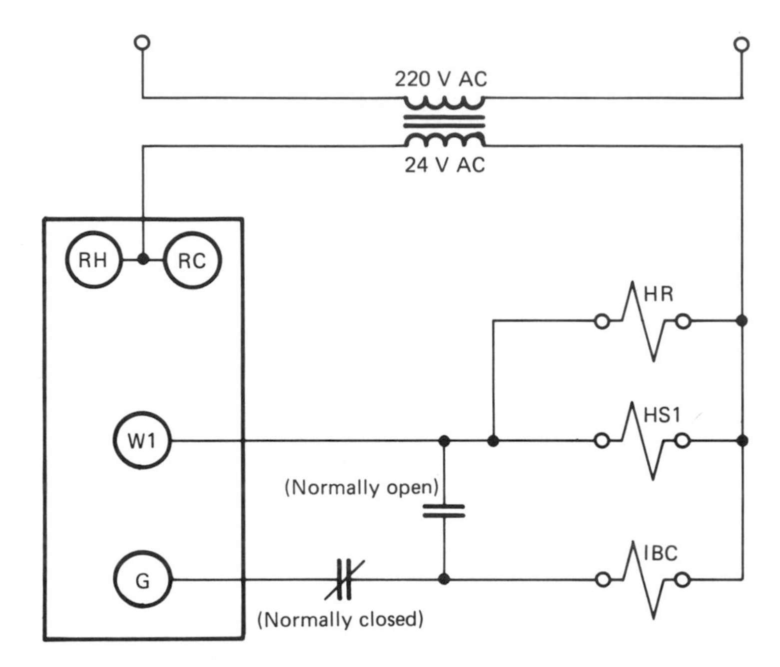

Figure 4–19. Addition of heat relay to provide positive control of IBC and heat strip 1 contactor during heating or cooling operation.

The circuit that has been developed is that of a relatively complete air-conditioning system as shown in Figure 4–20. If the addition of the components is followed in a step-by-step sequence from Figure 4–1 to the complete system in Figure 4–20, the overall system is easy to understand. The high-voltage circuits, as shown in final form in Figure 4–9, have been combined with the low-voltage circuits of Figure 4–19. The result (Figure 4–20) is a complete climate control system.

The circuit is completed with the addition of the fixed controls that are a part of most air-conditioning systems. They are:

1. Compressor 1 KLIXON (overload)
2. Compressor 2 KLIXON (overload)
3. Heat strip 1 overload
4. Heat strip 1 fuse link
5. Heat strip 2 overload
6. Heat strip 2 fuse link

Figure 4–21 shows the addition of these control components. (Note that overloads and fuse links may or may not show up in ladder diagrams.) The system of development described here should prove useful in determining what is required to control an air-conditioning system as well as to troubleshoot a malfunctioning system.

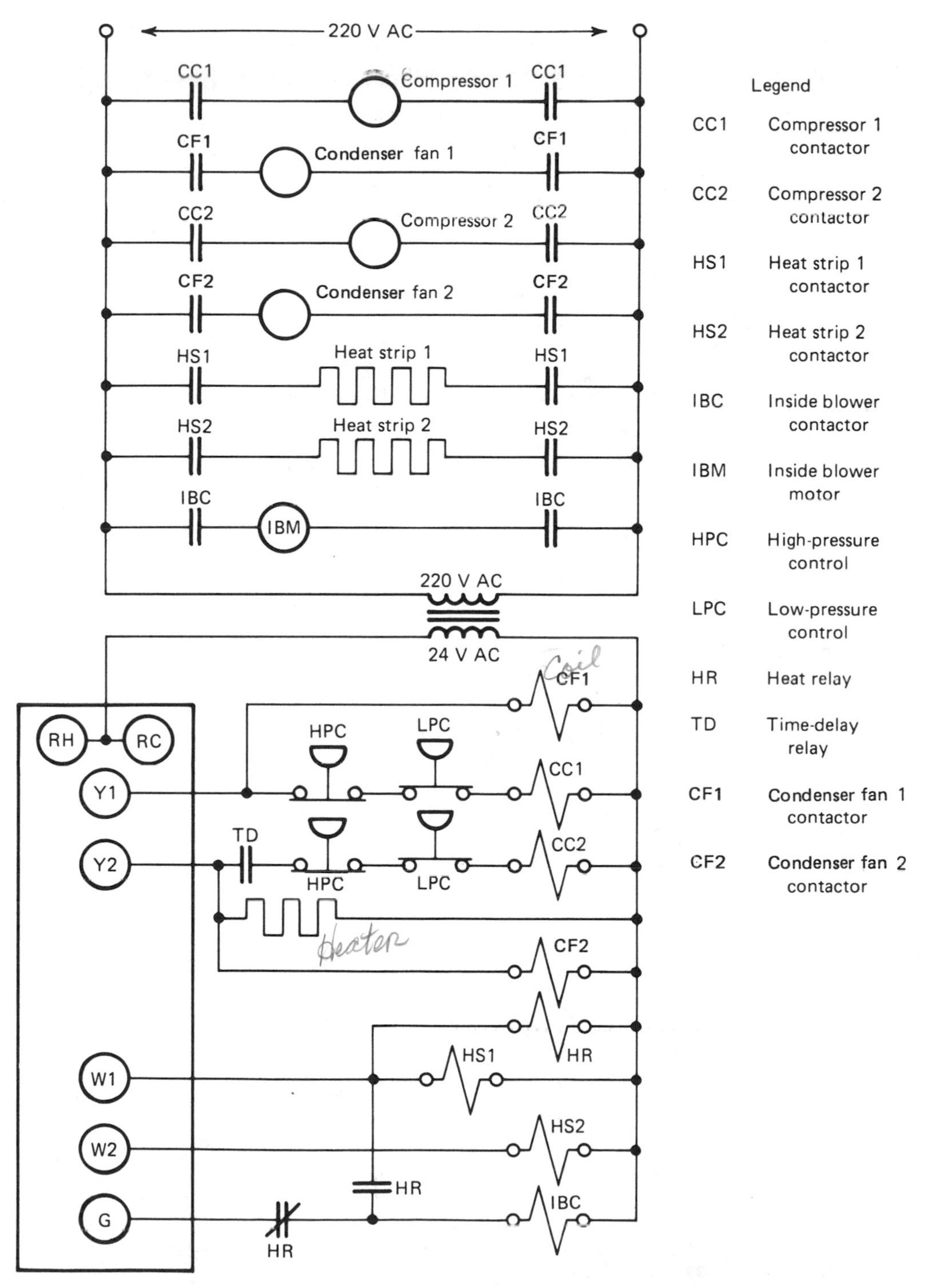

Figure 4–20. Ladder diagram of two-stage heat, two-stage cool air-conditioning system.

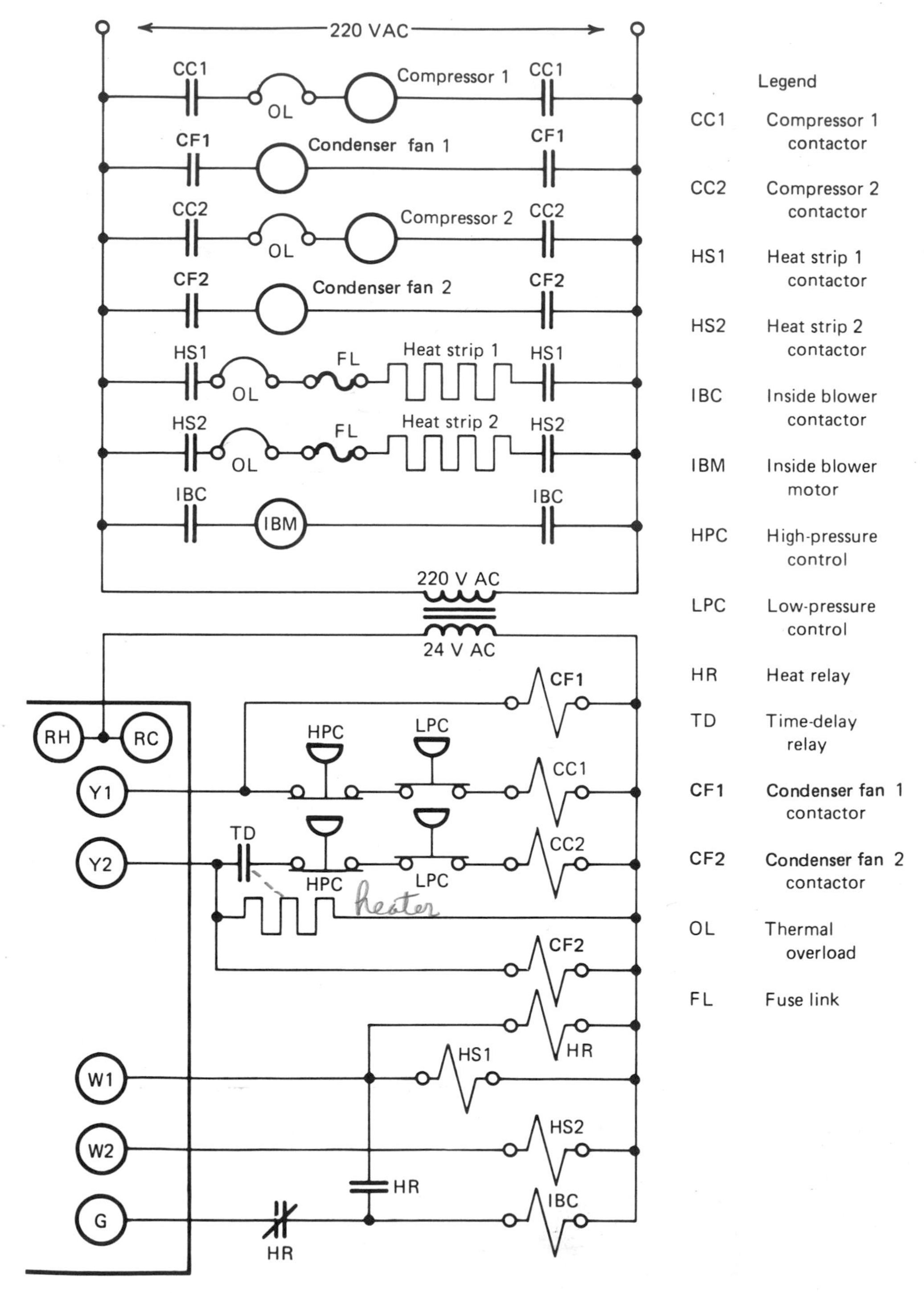

Figure 4–21. Ladder diagram of two-stage heat, two-stage cool air-conditioning system showing fixed controls.

THE REFRIGERATOR LADDER DIAGRAM

The ladder diagram of a refrigerator can be developed in the same manner that the air-conditioning ladder diagram was developed. The refrigerator is somewhat more simple since all components and controls operate directly from the 120-volt line.

Some of the components in the refrigerator circuit are connected directly across the 120-volt power source. These components operate continuously. The door heater and mullion heater are included in this group. They are shown in Figure 4–22 along with the power plug usually found in a home refrigerator. Whenever the power plug is in the wall socket, the door heater and mullion heater will be energized.

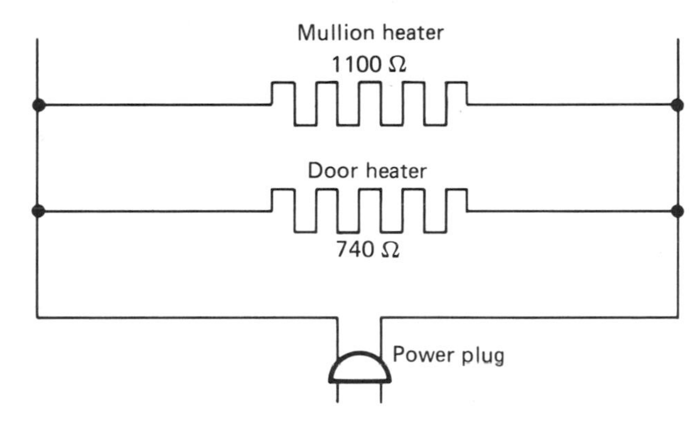

Figure 4–22. Refrigerator with continuously connected components.

There are usually both lights and fans in the refrigerator and freezer sections. The lights and fans are controlled by simple switches. The addition of these components is shown in Figure 4–23. When the door is closed, the refrigerator is on. When the door is open, the fan is off and the light is on. The freezer light is on when the door is open.

Now all that is needed for an effective refrigerator is an operating compressor and a thermostat (cold control). Since most refrigerators have a split-phase induction motor, a start relay will also be needed. The circuit, including the freezer fan and fan switch, is shown in Figure 4–24. Note that the switch that controls the freezer fan is located after the cold control. The switch is closed when the freezer door is closed. The freezer fan will operate only if the compressor is operating and the freezer door is closed.

Most modern refrigerators have a defrost system, which is an important part of the refrigerator system. The common defrost systems operate either on the preset operating time system or the accumulated time system. In the preset system, the defrost timer is operating continuously. This

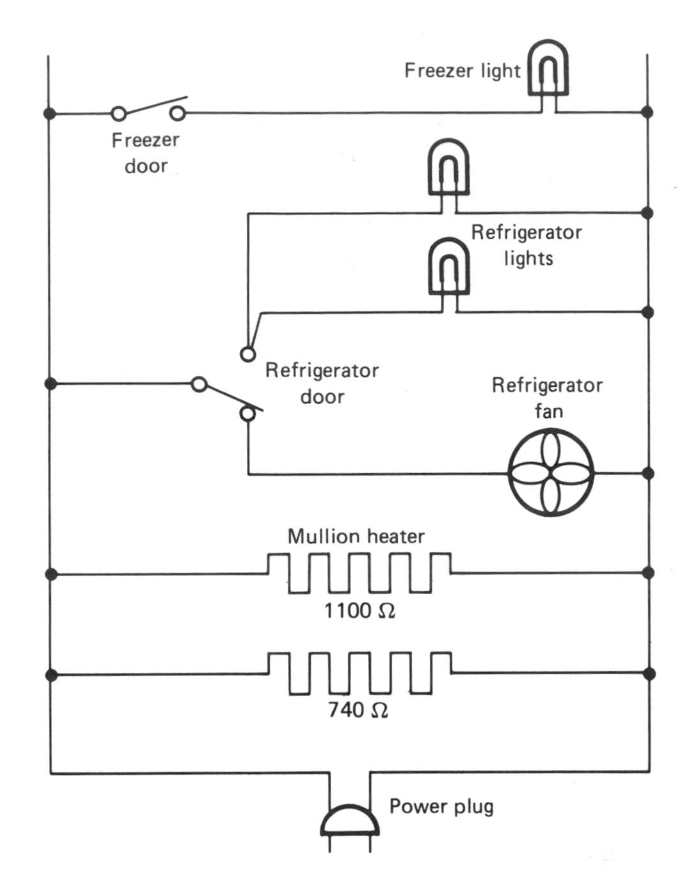

Figure 4–23. Addition of refrigerator fan, lights, and freezer light.

system is easily recognized by the location of the cold control (thermostat), as shown in Figure 4–25. If the cold control is located between the defrost timer and the start relay, the system defrosts with a preset timer controlled only by the timer motor and an associated cam switch. In Figure 4–26 the cold control is located before the defrost timer. This is the circuit of the accumulated time defrost system. The timer motor keeps track of how long the refrigerator is operating. When the cold control is closed, the timer motor and refrigerator operate. When the cold control opens, neither the refrigerator nor the defrost timer operates. This system provides an accurate indication of accumulation of refrigerator freezer moisture. The system in some measure responds to the number of refrigerator and freezer door openings.

Figure 4–26 is the ladder diagram of the accumulated time defrost system. In either system, a heater required for defrost control is connected

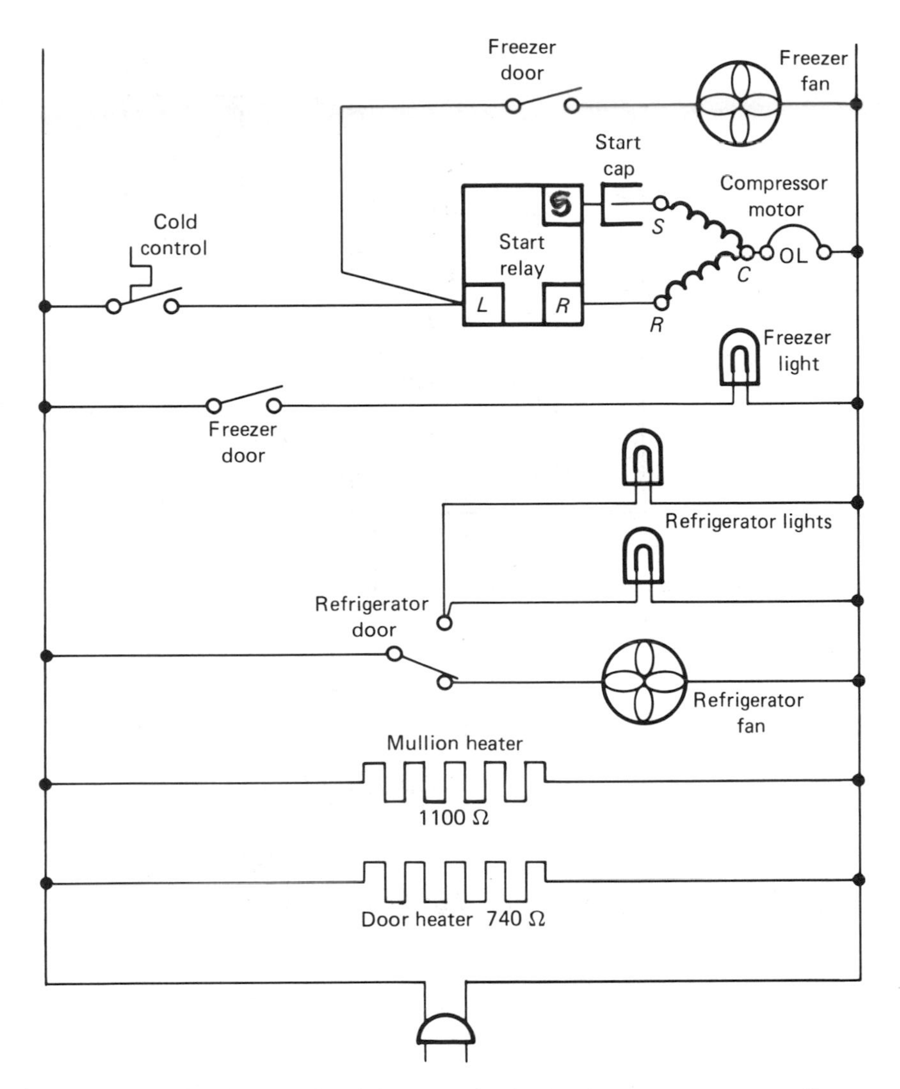

Figure 4–24. Addition of cold control, start relay, compressor, freezer fan, and fan switch.

directly across the power source. This defrost timer heater, along with the defrost timer, is shown in Figure 4–26.

The complete refrigerator-freezer system is shown in Figure 4–27. If each item is considered separately, the overall refrigerator-freezer system is relatively easy to understand.

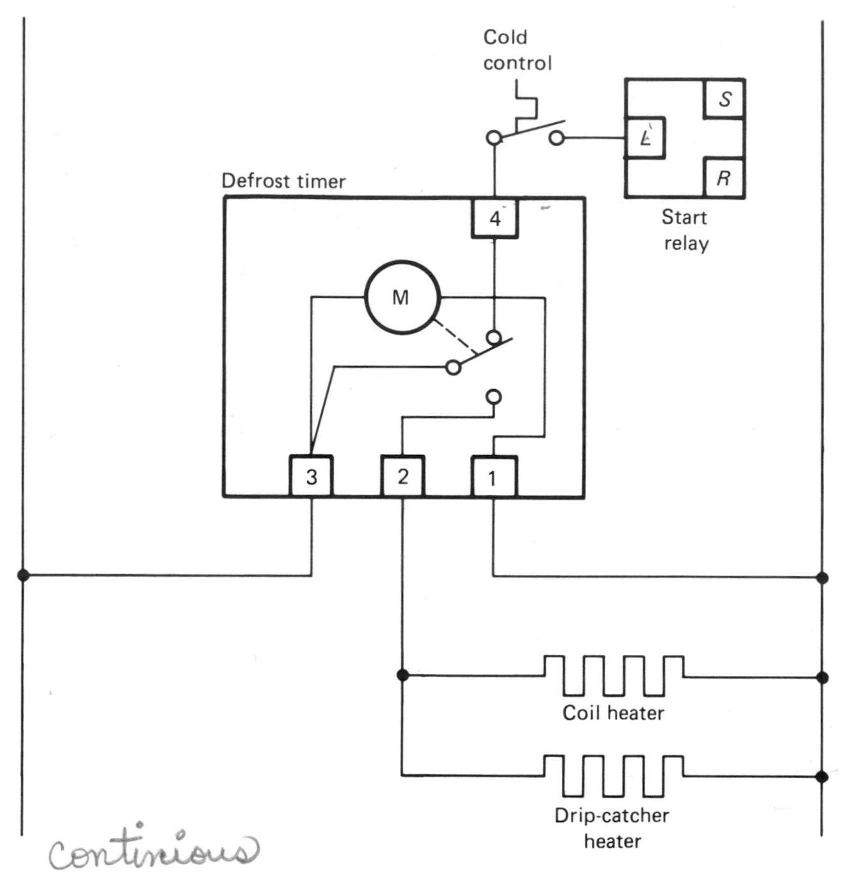

continious

Figure 4–25. Addition of defrost timer and heaters.

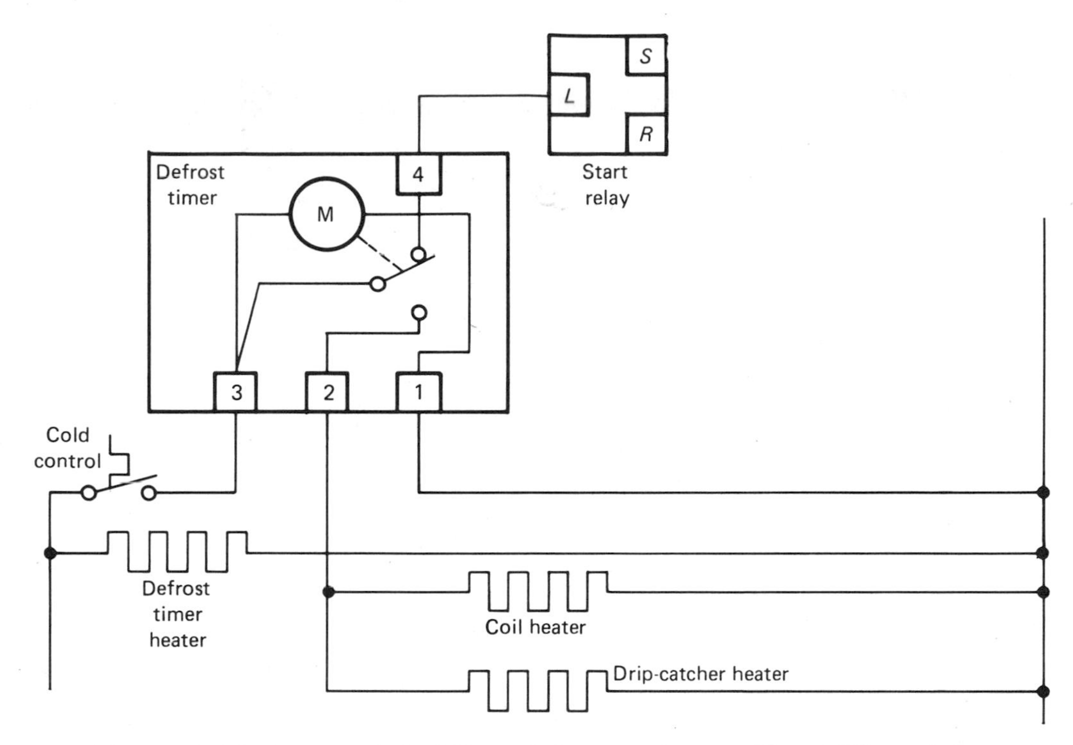

Figure 4–26. Accumulated time defrost systems with addition of defrost timer heater.

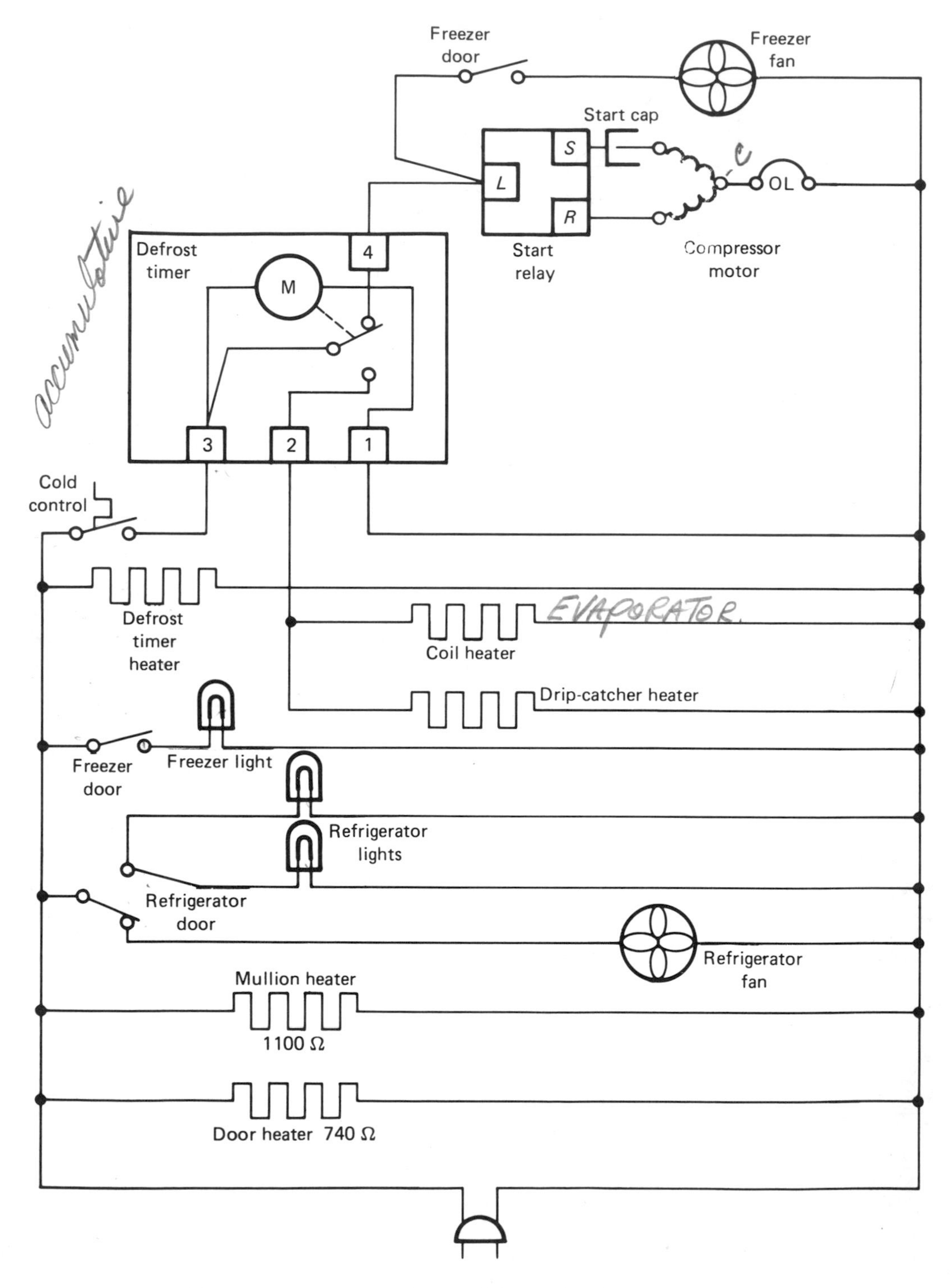

Figure 4–27. Complete refrigerator-freezer system.

Chapter 5

THERMOSTAT CIRCUIT DIAGRAMS

Thermostats come in many sizes, shapes, and forms. The circuitry inside the thermostats varies widely. When shown in a diagram, the thermostat usually consists of a block with input and output terminals, but little or no information is given on what goes on inside the block. For this reason, it is important that the technician make a general study of thermostats. When a thermostat or thermostat problem is encountered on the job, the operation of that particular unit should be investigated. A general approach to thermostats is taken here; description of a particular type or manufacturer is not intended. Whenever knowledge of a thermostat's operation is needed, the manufacturer's data sheet should be the first source of information.

When investigating a thermostat's operation, the technician frequently finds that the readily available information is in a form not easily understood. For example, the electrical drawing of a thermostat provided by the manufacturer might well be similar to Figure 5–1. This drawing is not too difficult to follow but does require a point-to-point tracing in order to ensure that actual circuits are considered. The drawing would have to be considered a combination pictorial schematic. The electrical circuits of the

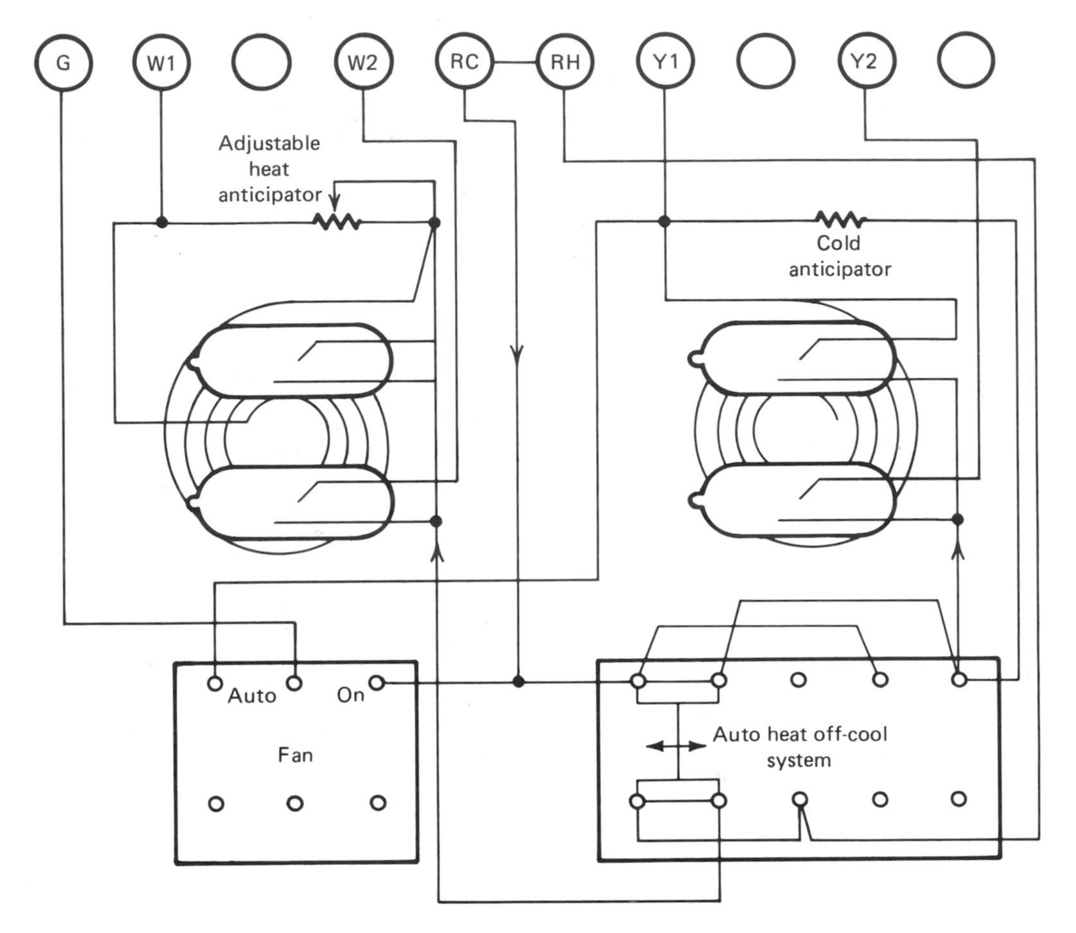

Figure 5–1. Thermostat of two-stage heat, two-stage cool air conditioner.

thermostat are more easily understood if a regular schematic diagram is prepared. Compare the schematic diagram of the same thermostat (Figure 5–13) with the pictorial schematic diagram (Figure 5–1). The schematic diagram is much easier to follow when electrical operation is the area of concern.

DEVELOPING THE HEAT-COOL THERMOSTAT

The most common element used in thermostats today is the bimetal strip. The bimetal strip is constructed of two metal sections formed into a single piece. One of the metal sections has a high expansion rate when exposed to heat, whereas the other section has a low expansion rate. The basic reactions of the bimetal strip are shown in Figure 5–2. The bimetal element is often

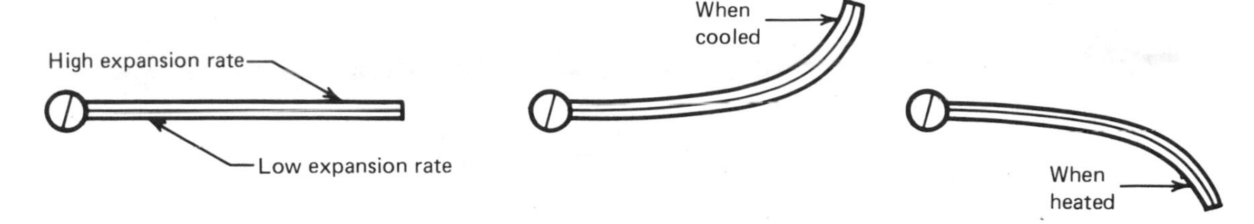

Figure 5–2. Bending of bimetal strip.

formed into a coil, which allows an increased movement at the strip end for small temperature variations, as shown in Figure 5–3.

The movement of the bimetal element can also be used as a means of making and breaking a circuit, as shown in Figure 5–4. Figure 5–5 shows how electrical circuits are used to control room temperature. As the bimetal strip cools, it will bend to the left, making an electrical connection with the fixed contact. The closed electrical circuit may be used to turn on a heating unit. As the air in the room and especially around the thermostat heats, the

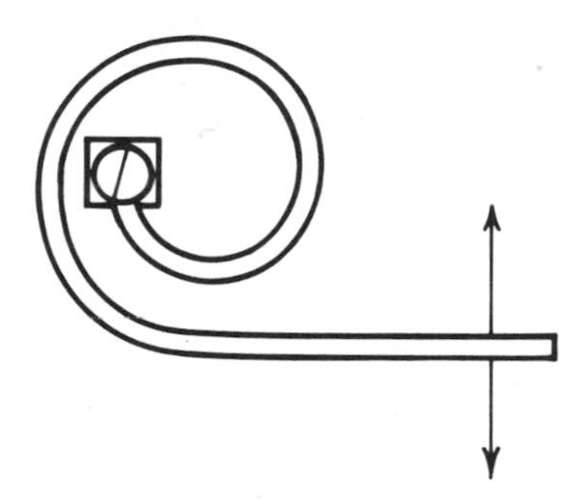

Figure 5–3. Coiled bimetal strip allows increased movement for small temperature variations.

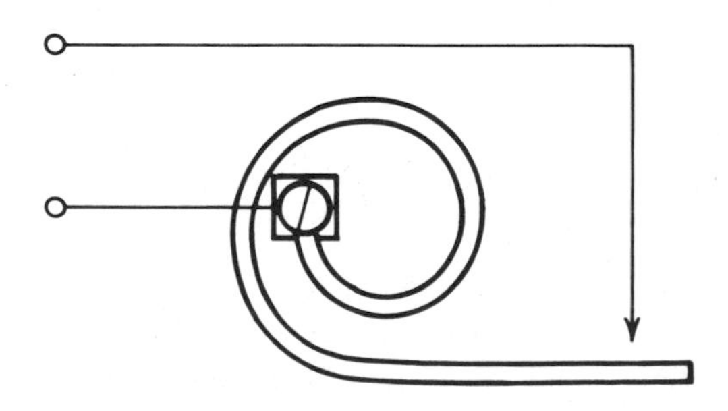

Figure 5–4. Use of bimetal strip to make electrical contacts.

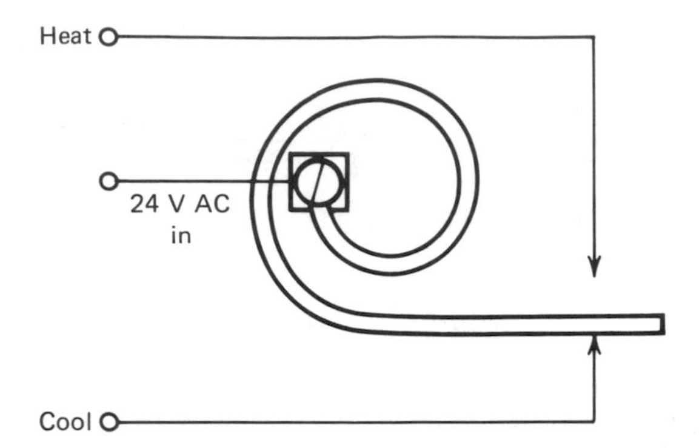

Figure 5–5. Thermostat with heat and cool outputs.

bimetal strip will bend away from the contact, finally opening the circuit and causing the heating unit to be turned off. A cooling thermostat can be provided by placing another contact on the opposite (right) side of the bimetal strip. Heat will cause the bimetal to move toward this contact. The electrical circuit through this contact is used to turn on the cooling unit.

Switches are provided in the output lines to ensure that the heating and or cooling units are turned on only when needed—that is, that the cooling

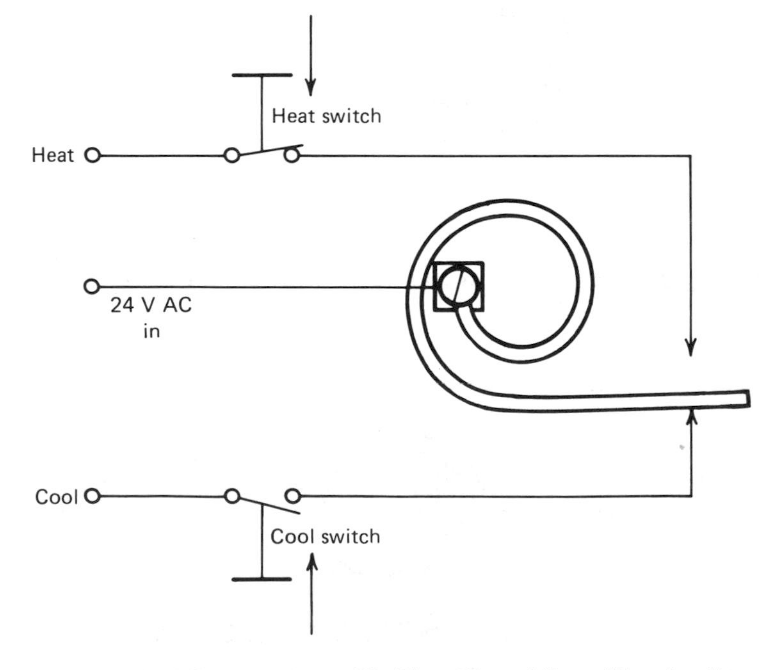

Figure 5–6. Thermostat with "heat" and "cool" selection.

unit will not come on when "heat" is selected and that the heating unit will not come on when "cool" is selected. The addition of the switches is shown in Figure 5–6. When the heat selection button is pressed, the switch in the heat output circuit will close and the switch in the cool output circuit will open. When the cool selection button is pressed, the switch in the cool output circuit will close and the switch in the heat output circuit will open.

ANTICIPATORS

Anticipators are used in thermostats to provide for smaller variations in room temperature from the selected temperature. Consider the situation where "heat" is selected, and the thermostat is set at the selected temperature. When the room cools to below the selected temperature, the heat contacts in the thermostat will close. The heating unit will begin to operate. Before heat can be brought into the room, the temperature in the heating unit must rise. During this time, the room or area to be heated continues to cool down. When the temperature in the heating unit finally reaches the proper level, heat will be provided to the selected area. The thermostat will sense the selected temperature when the room reaches that temperature. The thermostat will open, shutting down the heating unit. The heat that is still in the heating unit will continue to be delivered to the room, causing the temperature to rise above the selected temperature. The straight bimetal

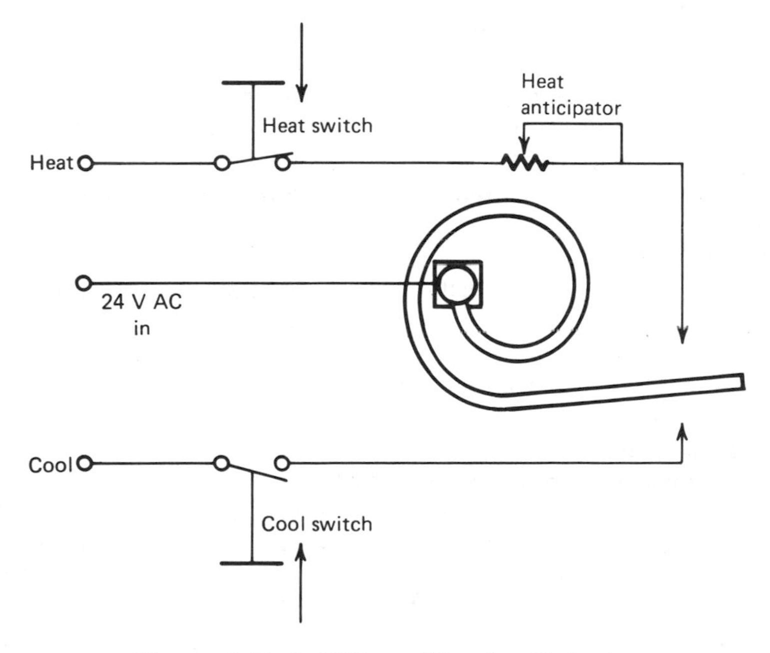

Figure 5–7. Addition of heat anticipator.

thermostat will control room temperature, but the room temperature will vary widely from the selected temperature.

Heat Anticipators:

A heat anticipator is an adjustible resistor that provides heat inside the thermostat whenever the contacts are closed. The addition of a heat anticipator to the thermostat will decrease the magnitude of the variation in temperature from the selected value. Figure 5–7 shows the connection of a heat anticipator in a thermostat. The anticipator will provide heat in proportion to the amount of current drawn by the heat valve or relay that is controlled by the thermostat. The adjustable anticipator (see Figure 5–8) is

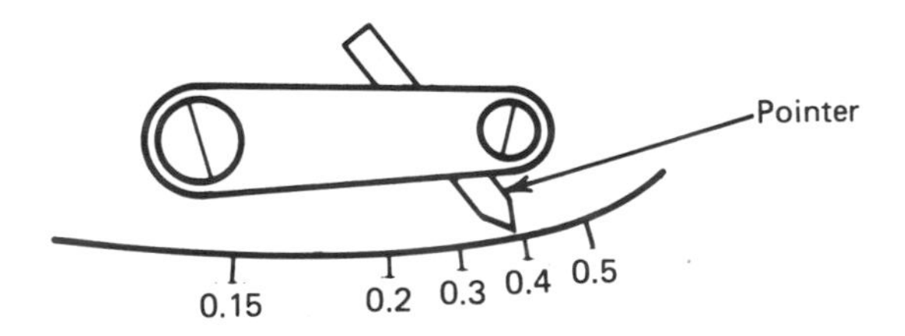

Figure 5–8. Heat anticipator.

normally set with the pointer indicating the draw of current of the external valve or relay. The adjustment is shown set at 0.4 amps, a common heat setting.

The heat anticipator adjustment will control the "on" period of the heater. A higher setting of the anticipator will increase the "on" time and decrease the cycling frequency of operation. A lower setting of the anticipator will decrease the "on" time and increase the cycling frequency.

Cool Anticipators:

Cool anticipators perform a function similar to that of heat anticipators with regard to keeping room temperature constant. The connection of the cool anticipator is shown in Figure 5–9. Since the cool anticipator resistor is connected across the open cool contacts, the full control voltage of the thermostat is across the anticipator resistor whenever the thermostat cool contacts are open.

The function of the cool anticipator is to provide heat inside the thermostat when "cool" is selected (the "cool" button is pressed) and the thermostat is open. This will cause the thermostat's cool contacts to close sooner. As the contacts close, the cool anticipator resistor is shorted out, as shown in the inset in Figure 5–9. Further heat is not produced in the anticipator resistor, and the thermostat cools normally. When the room cools to the selected temperature, the thermostat will open, turning off the air

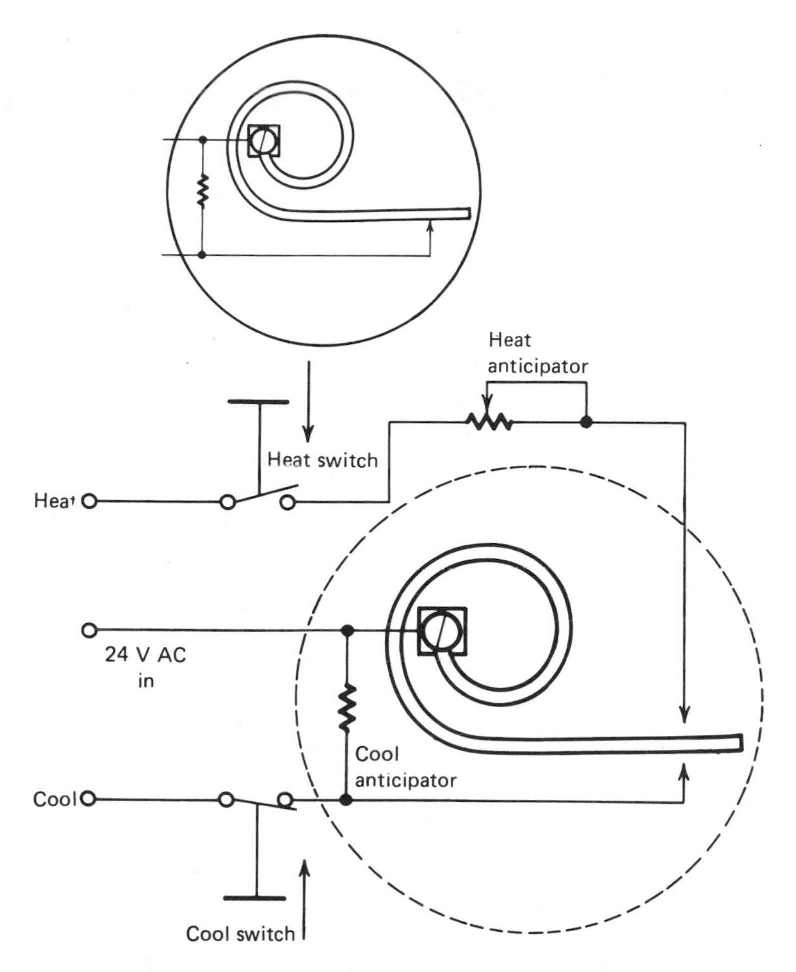

Figure 5–9. Addition of cool anticipator.

conditioner. The cool anticipator resistor is then reconnected in the circuit and will start to heat up again. This heat will again cause the contacts to close a little sooner than they would have without the anticipator resistor heat. The use of the cool anticipator will result in an increase in frequency of "on"/"off" cycling of the air conditioner and will therefore provide for a smaller variation in room or controlled-area temperature.

FAN CONTROL

The indoor fan or blower is another subsystem that is controlled at the thermostat. The fan switch is usually a two-position switch with "on" and

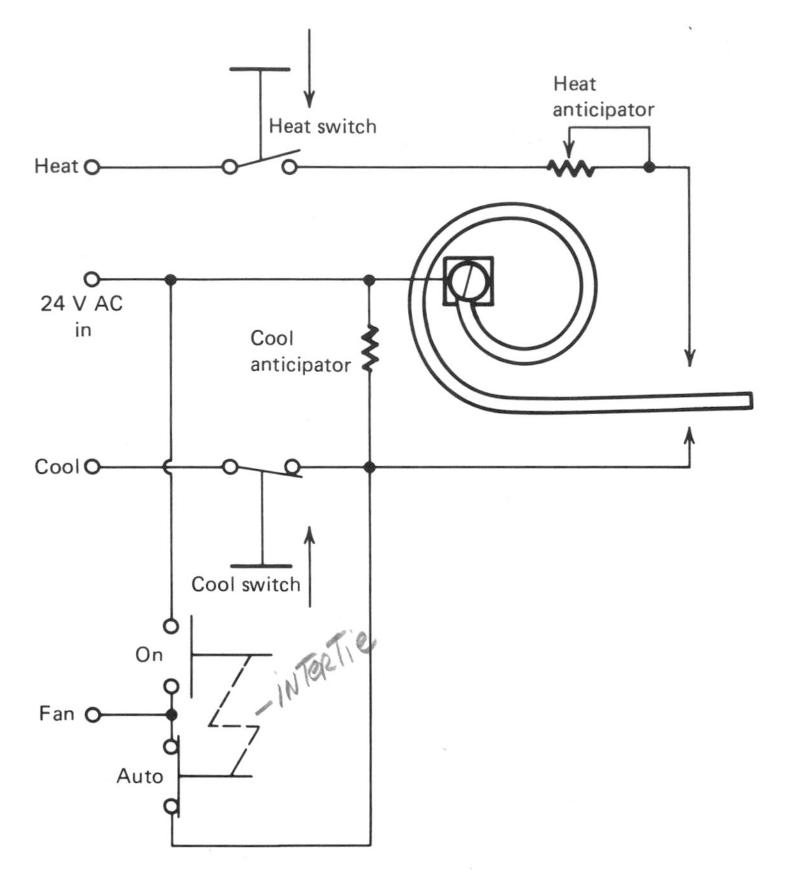

Figure 5–10. Addition of fan switch.

"auto" as the switch positions. The thermostat is a complete unit after the addition of the "fan" switch, as shown in Figure 5–10.

There is another condition that must be considered when thermostats operate in heating and/or cooling situations. It has already been shown that a heat relay is needed when the same inside blower is used for heating and cooling (refer to Chapter 4, "Heat Strip Contactor Control"). The connection of the heat relay will cause additional current to be drawn through the heat anticipator resistor. Depending on the amount of the additional current, sufficient heat may be produced in the anticipator resistor to cause premature openings of the thermostat contacts and therefore premature cycling off of the heat source.

Whenever a thermostat with a heat anticipator is used, the total current drawn through the anticipator resistor must be considered and the anticipator resistor should be adjusted according to that level of current.

Check the gas valve for rating

MERCURY BULBS

Any switch contacts that are open to the air will oxidize and will also pit slightly each time they are opened. A mercury switch can effectively eliminate this pitting and oxidizing. Figure 5–11 shows a mercury switch mounted on a bimetal thermostat.

When the requirement for limiting the variations in temperature became more important, the single bimetal strips used for heat and cool were changed to double strips in some thermostats. Figure 5–1 shows a double bimetal strip thermostat with one section used for heat and the other section used for cool.

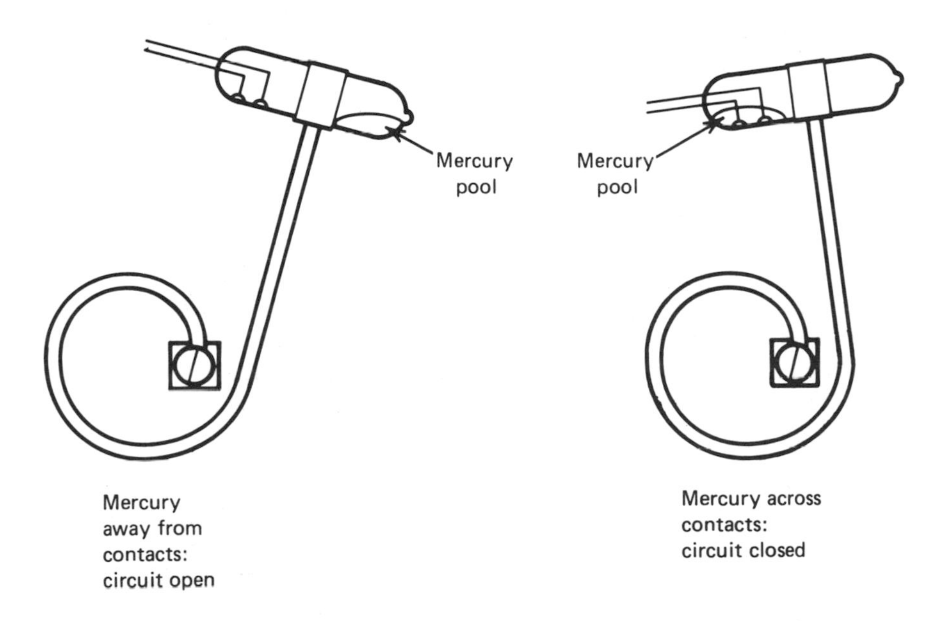

Figure 5–11. Bimetal thermostat with mercury switch.

DEVELOPING THERMOSTAT SCHEMATICS

The operation of the thermostat shown in Figure 5–10 is easy to follow. But even this diagram requires that the circuit be traced. Consider how much simpler that same circuit is to follow when it is presented in schematic form as in Figure 5–12. The two-stage heat, two-stage cool thermostat shown in Figure 5–12 is the same thermostat that is shown in Figure 5–1. As mentioned earlier, it is much easier to follow the circuit in the schematic diagram (Figure 5–13) than it is to trace the circuit in Figure 5–1.

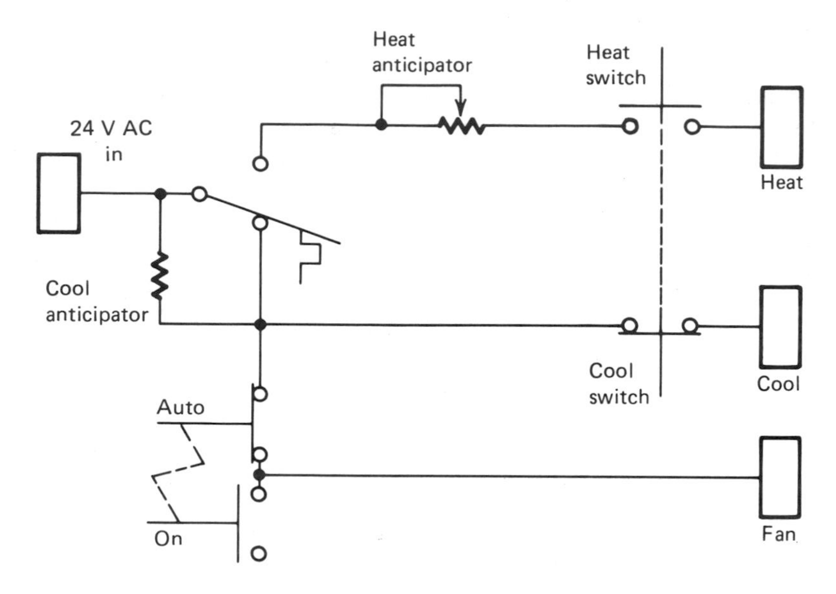

Figure 5–12. Schematic diagram of simple thermostat.

There is a present-day trend toward using a sub-base with low-voltage thermostats. The sub-base functions as a mounting plate for the thermostat. Miniature switches for circuit connections are often included in the subbase. The manufacturer's manual or data sheet should always be referred to when it is necessary to investigate the thermostat circuitry.

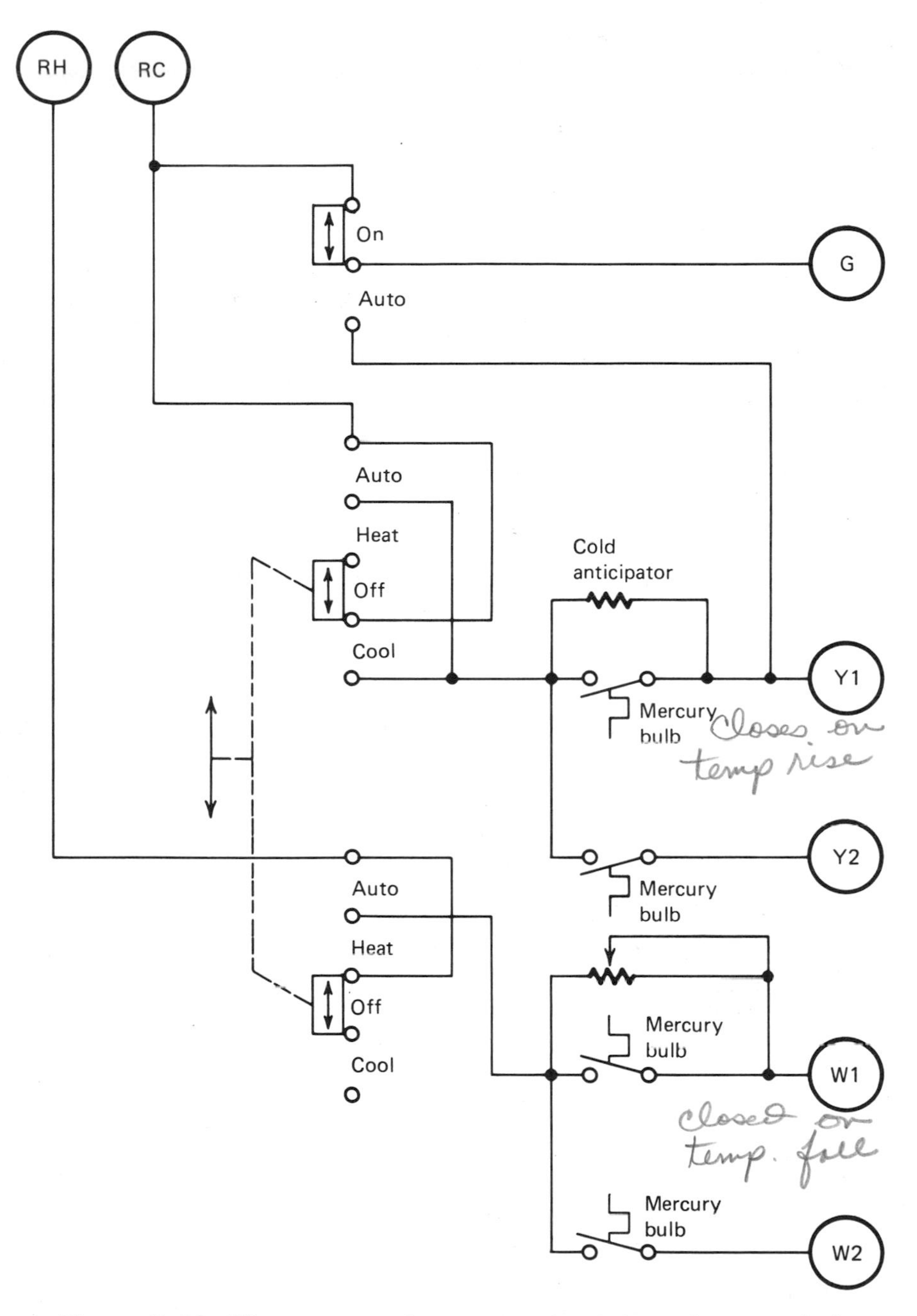

Figure 5–13. Thermostat of two-stage heat, two-stage cool air oonditioner.

Review Questions

Chapter 5

Mark each statement T for true or F for false.

1. A bimetal element bends when heated because the metal softens with heat. 1. F
2. A single bimetal element may be used for both heat and cool control in a thermostat. 2. F
3. The cool anticipator resistor must be adjusted to match the cool relay. 3. __
4. The heat anticipator resistor adjustment should be set to match the total current drawn through it. 4. T
5. Moving the anticipator adjustment in Figure 5–8 from 0.4 amps to 0.2 amps will increase the cycling frequency. 5. T
6. The cool anticipator resistor is shorted out when the cool contacts mate. 6. T
7. The addition of a heat relay to the thermostat heat output circuit may call for an adjustment of the heat anticipator resistor. 7. T
8. Mercury bulbs are used in thermostats because they are less expensive than open contacts. 8. F
9. Two problems are eliminated when mercury bulbs are used in place of open contact: pitting and oxidation. 9. T
10. Schematic diagrams of thermostats are easier to follow than pictorial diagrams. 10. T

Chapter 6

THE WIRING DIAGRAM

The wiring diagram and the ladder diagram are both important ways of picturing the circuits of air-conditioning systems. Just as the ladder diagram was developed in Chapter 4, the wiring diagram will be developed in this chapter. The final circuit will be an air-conditioning system using a three-phase compressor for cooling and strip heat for heating. Control circuits for the system will be supplied with 24 volts AC. When high-voltage and low-voltage circuits are shown on the same drawing, the high-voltage circuits will be shown with thick lines and the low-voltage circuits will be shown with thin lines.

HIGH-VOLTAGE CIRCUITS

In Figure 6–1, the circuit of the three-phase, 208-volt input is shown connected to the main disconnect switch. The output of the disconnect switch is fed directly to the compressor contactor terminals.

POINT OF SAFETY
The main disconnect switch is normally mounted on the **condenser** *unit or in an area close by the* **condenser**. *The purpose of the disconnect switch is to remove power from the unit. It is important to*

note that when the disconnect switch is located at the **condenser**, *the technician servicing the equipment has control over the application of power to the equipment. If the disconnect switch is located away from the* **condenser**, *someone not knowing that a technician is working on the equipment might reapply power. The closer the disconnect is to the equipment, the safer the situation.*

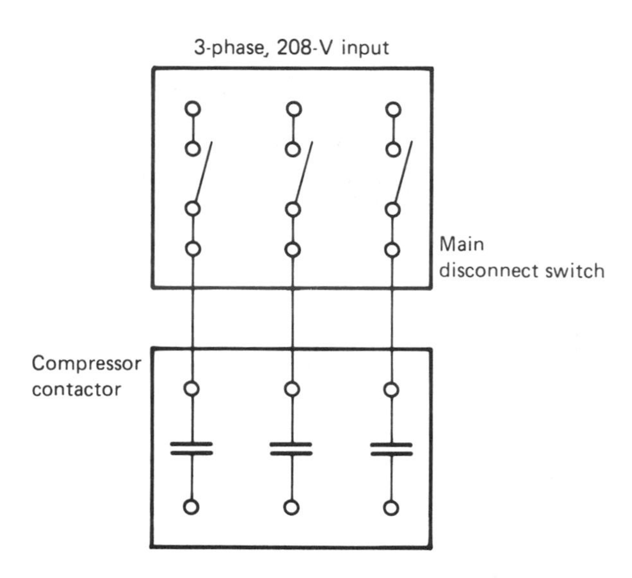

Figure 6–1. Main power input (three-phase).

When the compressor contactor is energized, power is fed through the closed contacts to the three-phase compressor motor. The compressor motor circuit includes two current overload protectors in the high-voltage circuits. The completion of three-phase power connections to the compressor is shown in Figure 6–2.

If 208 volts of three-phase power is available at the disconnect switch, the disconnect switch closes and the compressor contactor is energized; the compressor motor should run.

The line current overload protectors are a combination of a snap disk and a set of contacts actuated by the snap disk. If the motor should draw too much current, the disk will snap, opening the contacts associated with it. The overload situation is shown in detail in Figure 6–3.

The other circuits requiring the application of high voltage are the outside blower motor, the inside blower motor, and the control transformer. These circuits operate on 208-volt single phase power. Follow the circuits in Figure 6–4, which shows the addition of the outside blower motor. The

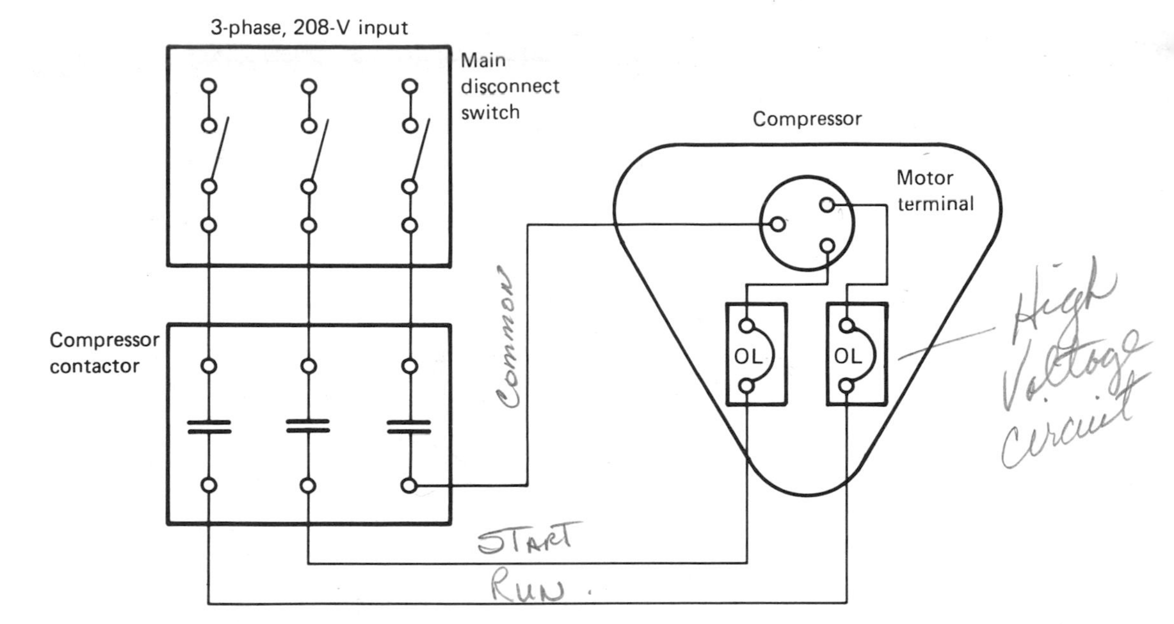

Figure 6–2. Connection of main power input to compressor.

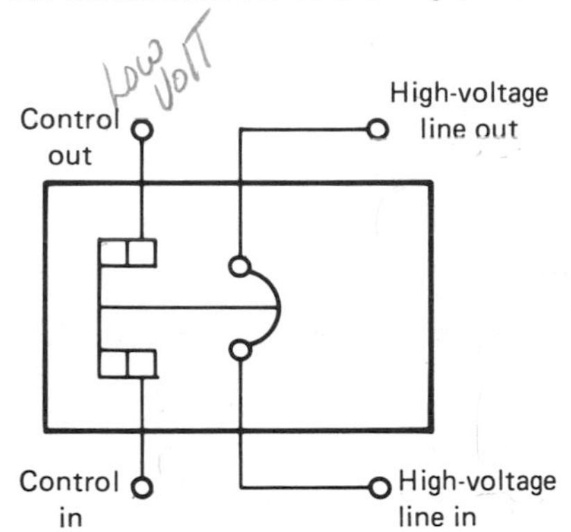

Figure 6–3. Overload.

outside blower motor is a single-phase capacitor motor. Power is fed from the output side of the compressor contactor to the outside blower motor.

The control transformer primary is connected to the input side of the compressor contactor. Terminal strips are included in order to provide convenient tie points for the high-voltage connections. The additional components are shown in Figure 6–5.

The last circuit associated with the air conditioner requiring high-voltage power is the inside blower motor. This motor may operate when the

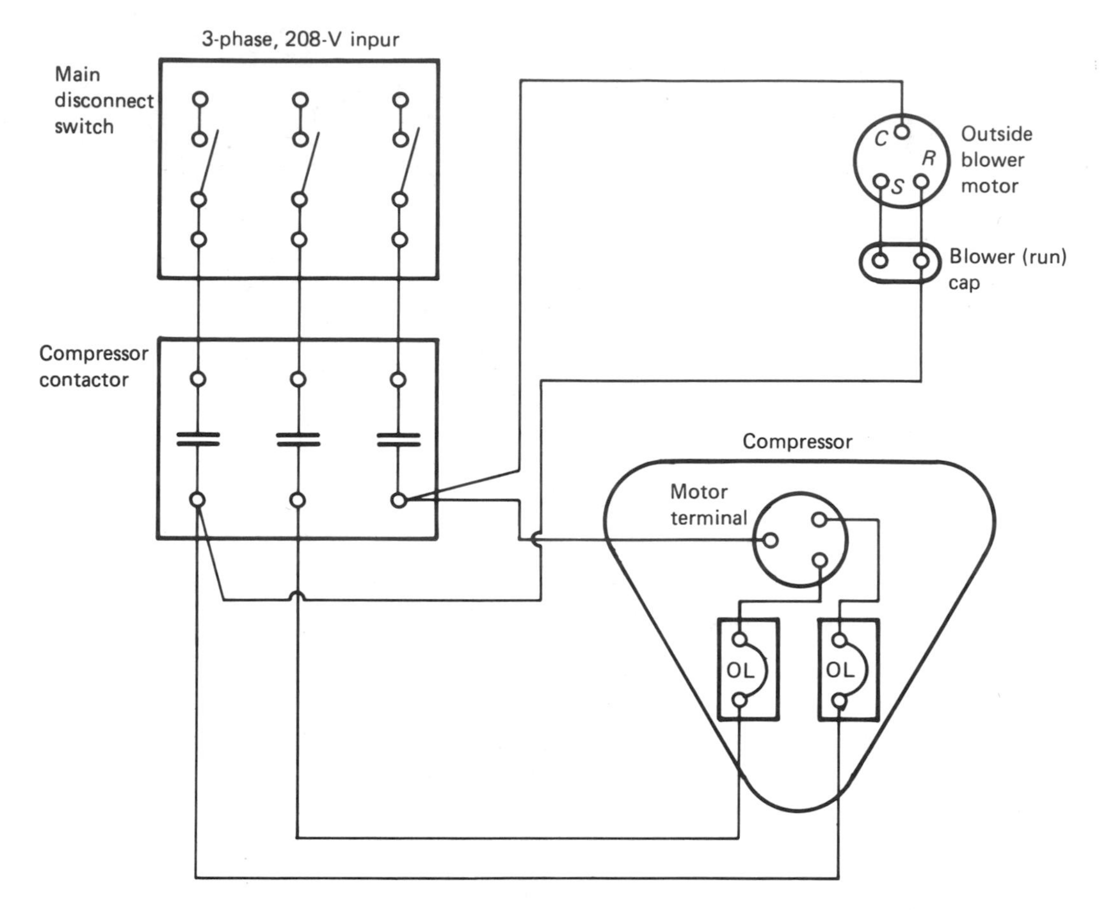

Figure 6–4. Addition of outside blower motor.

compressor contactor is off. The power for this motor is taken from the input side of the compressor contactor.

The purpose of the terminal strips is to limit the number of connections at any one terminal site. For example, there are already two connections at some of the compressor contactor terminals. Two should be the limit under any one screw connection. The terminal strips provide additional connection sites for the same electrical points. Figure 6–5 shows the connections of the inside blower motor and the blower relay contacts. The addition of the inside blower motor completes the high-voltage circuit of the air conditioner.

LOW-VOLTAGE CIRCUITS

The primary of the control transformer (low-voltage transformer) was fed directly from the input side of the compressor contactor. Power will be fed to

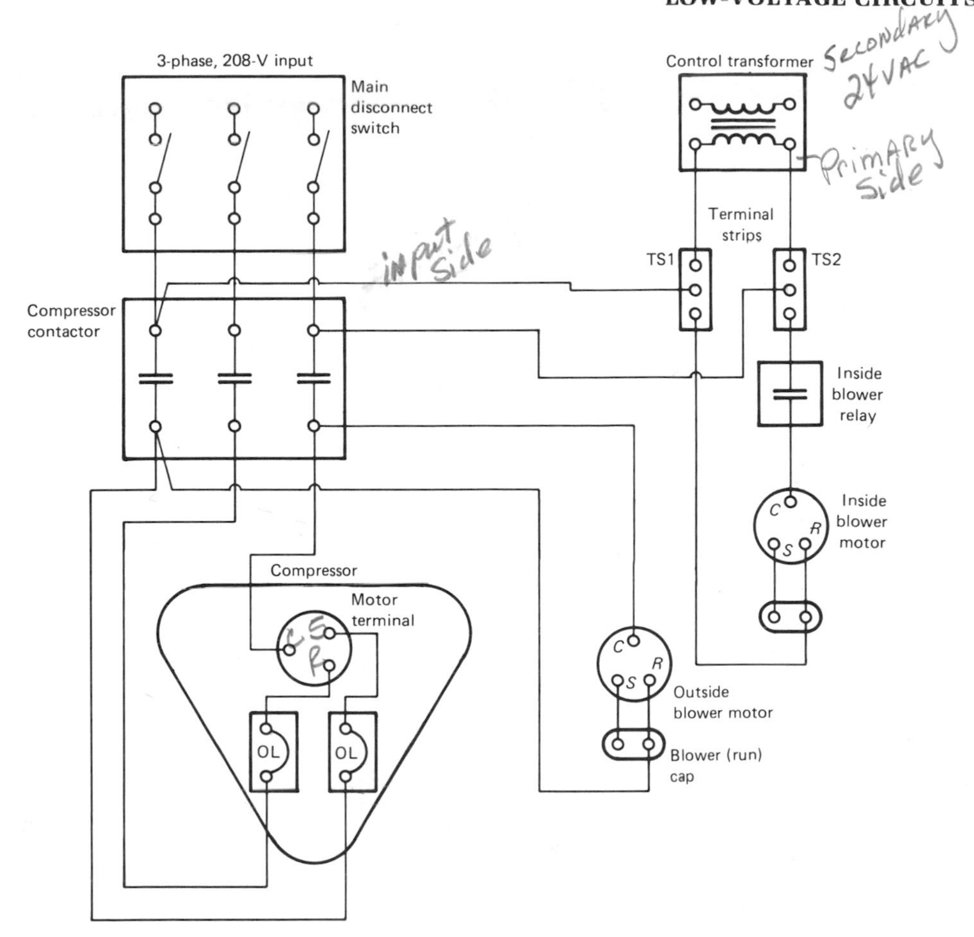

Figure 6–5. Complete cooling high-voltage circuit.

the primary of the transformer whenever the disconnect switch is closed.

The compressor low-voltage circuits consist of the thermostat anti-short-cycle device, the compressor contactor, and the fan relay. The low-voltage circuits will be developed first, and the complete compressor system of high- and low-voltage circuits will then be presented.

The secondary of the control transformer has 24 volts AC across its terminals. The return side of the transformer is connected directly to terminal strip TS3 in Figure 6–6. The needed connection points are available at TS-3. The red lead from the control transformer is fed directly to the *R* input of the thermostat. When the temperature of the air surrounding the thermostat is above the thermostat setting, the 24 volts AC will be available at the *Y*1 output of the thermostat.

The first connection of the *Y*1 or cool output of the thermostat is to the anti-short-cycle device, which was recently developed for compressor

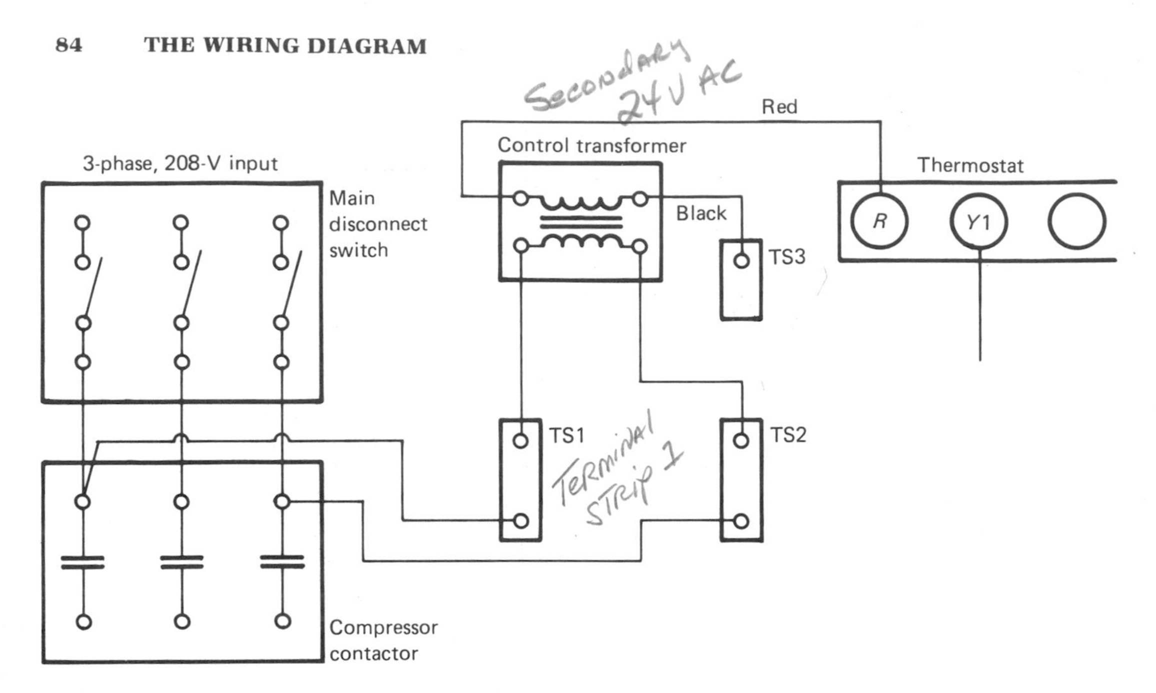

Figure 6–6. Low voltage control transformer circuit.

protection. Should electrical power be interrupted for any reason, the anti-short-cycle device will prevent the reapplication of power until after the compressor high-side pressure has reduced to a low level. The time delay is approximately three minutes. This particular unit will also interrupt the 24-volt circuit to the compressor contactor coil if a "brownout" condition (low input, live voltage) exists. The compressor motor cannot effectively start against high-side pressure; nor should operation continue against high-side pressure if the supply of voltage is low. Both are burnout conditions. The connection of the anti-short-cycle device is shown in Figure 6–7. Under normal conditions, a circuit exists between input terminal 1 and output terminal 3 of the device. When the device is activated, that circuit is open.

The compressor low-voltage circuit continues from the anti-short-cycle device to the overload protectors located at the compressor. The overload protectors are controlled by two lines of the three-phase input lines. The secondary contacts of these overload protectors are used in the low-voltage control circuit. The contacts are connected in series.

Another protection device for the compressor is the thermal overload. The thermal overload is built into the compressor in order to pick up internal heat. The contacts of the thermal overload snap open when the internal temperature of the compressor exceeds the compressor manufacturer's preset limit. The thermal overload is connected in series in the low-voltage control circuit.

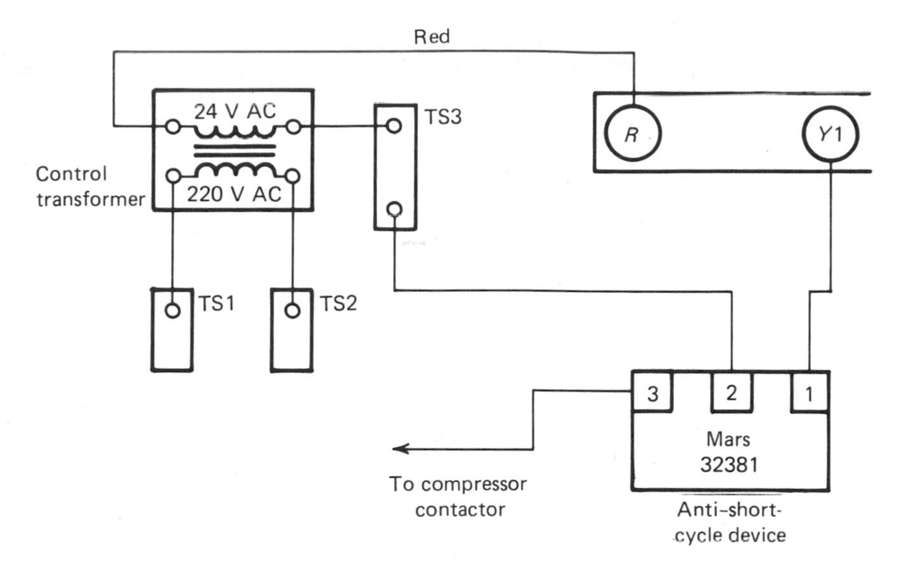

Figure 6–7. Addition of anti-short-cycle device.

The last protective device connected in the control circuit to be discussed here is the dual-pressure control. The dual-pressure control senses a problem in the refrigerant system that could cause further problems for the compressor. If the pressure is too high, it usually means a blocked system. If the pressure is too low, there is usually a leak in the system. It is appropriate to keep the compressor from operating under too high or too low a pressure. One of the controls will open a set of contacts in series with the control circuit if the pressure is above or below the preset limit. The remainder of the compressor low-voltage control circuit is shown in Figure 6–8. As it is described in the following paragraph, trace the circuit in Figure 6–8.

The low-voltage control circuit is a series circuit. Starting at the control transformer red lead, follow the circuit to the thermostat. When the thermostat calls for cooling, the circuit is completed to the *Y*1 output. The yellow wire completes the circuit from the thermostat to the terminal 1 input of the anti-short-cycle device. If a low-voltage (brownout) condition does not exist, and at least three (3) minutes have passed since the compressor was last turned on, a circuit will exist through the anti-short-cycle device from terminal 1 to terminal 3.

Follow the circuit from terminal 3 of the anti-short-cycle device to the thermal overload on the compressor. The circuit continues to overload protector 1 on terminal 3. The circuit is complete through overload protector 1, 3, and 4 terminals and overload protector 2, 3, and 4 terminals. The circuit

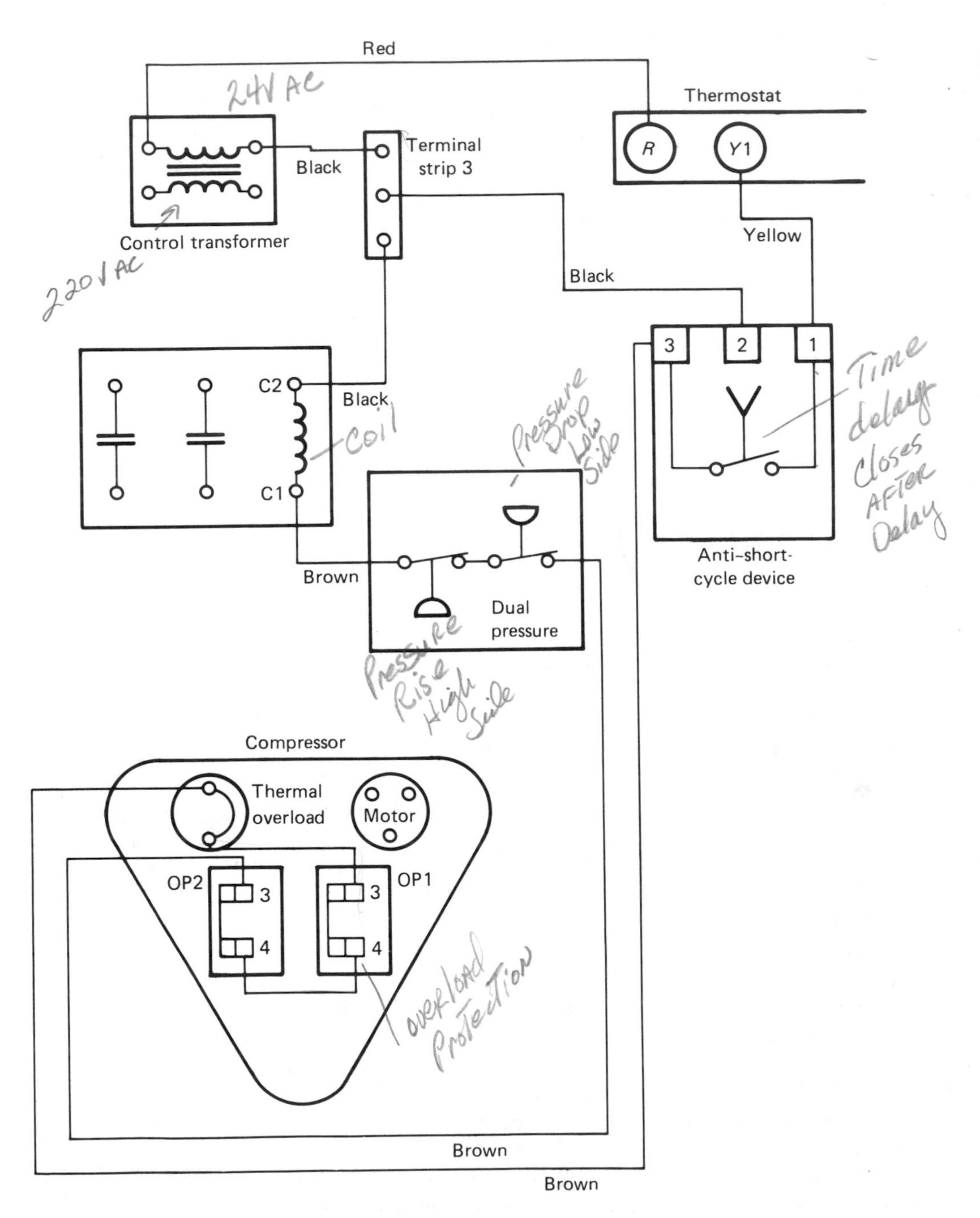

Figure 6–8. Compressor low-voltage control circuit.

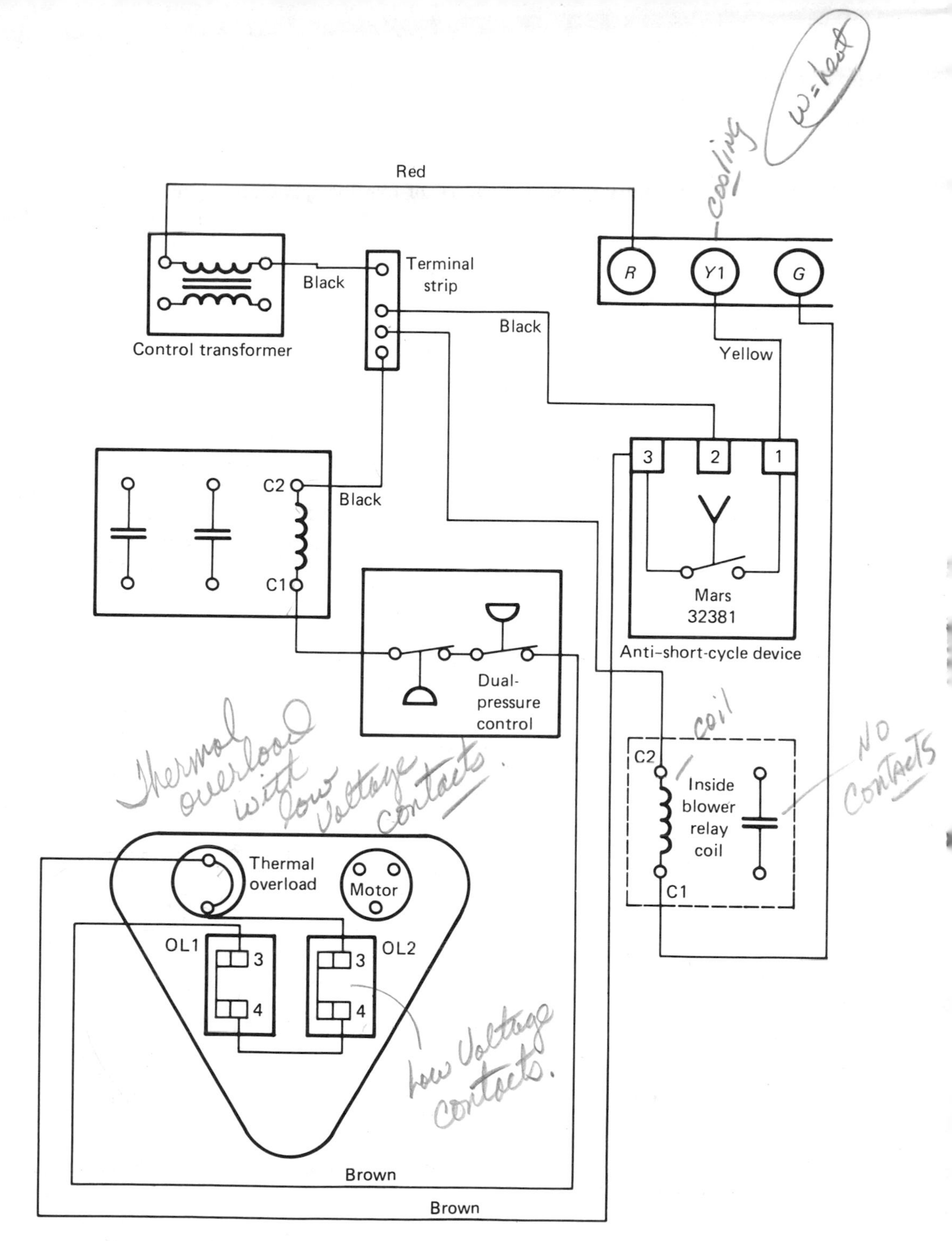

Figure 6–9. Addition of inside blower relay coil.

continues from the 3 terminal of overloal protector 2 to the thermal overload. From the thermal overload, the path continues through the dual-pressure control to the contactor terminal C1. The return path is from the contactor coil terminal C2 to terminal strip TS3, which is directly connected to the control transformer secondary through the black wire.

The final addition to the low-voltage control circuit is the connection of the inside blower relay coil. The circuit is added in Figure 6–9. When the thermostat calls for operation of the inside blower motor, a circuit will be completed through the thermostat from *R* to *G*, completing the path to terminal C1 of the inside blower relay coil. Terminal C2 of the coil has a complete path to TS3 and the return side of the low-voltage transformer.

In order to locate the actual connection point of wires, it is necessary to refer to a wiring diagram. Consider, though, how much easier it is to follow circuit operation when a ladder diagram is provided. In Figure 6–10, the low-voltage circuit is presented in ladder form. Even though the ladder diagram

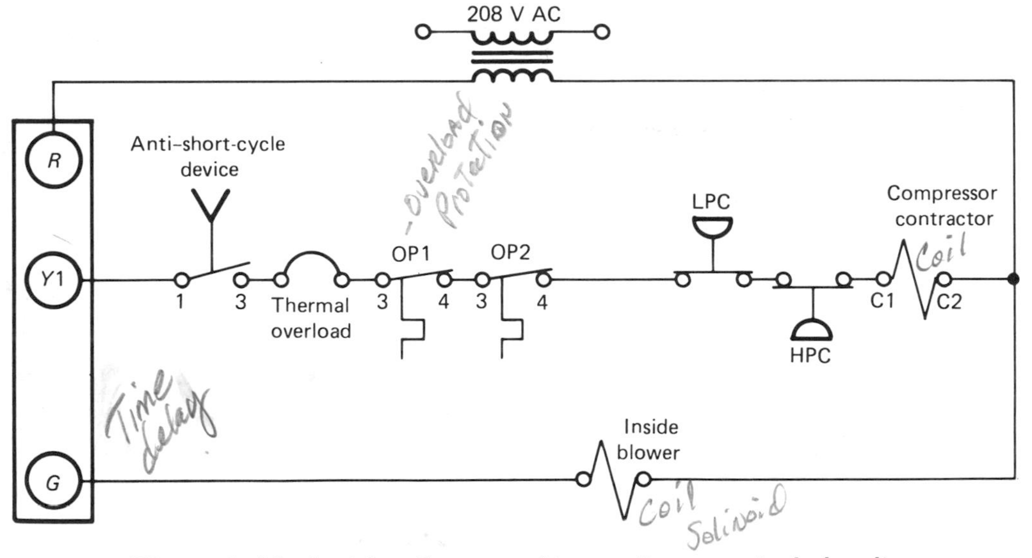

Figure 6–10. Ladder diagram of low-voltage control circuit.

is easier to follow, it does not present enough information to ensure effective troubleshooting.

HIGH-VOLTAGE HEAT CIRCUITS

Heat units 1 and 2 use the same type circuit. Heat is generated in a resistance element when voltage is applied to that element. Protective devices are

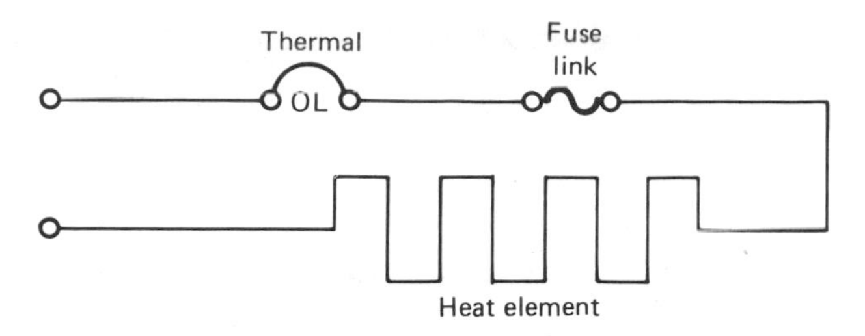

Figure 6–11. Electric heat unit.

provided in the form of an overload and a fuse link. Figure 6–11 is a diagram of the heat unit.

The overload is a thermal-type device that opens if the heat unit becomes too hot. If the overload should fail to open at the preset limit, the fuse link will melt a short time after the preset limit is reached. Either the overload or the fuse link will open the circuit to the resistive heater. When the current flow through the heat unit is interrupted, it will cool down.

The high voltage to the heat unit is supplied directly from the main disconnect switch. These connections are shown in Figure 6–12. The heat strip is a single-phase device; therefore, only two of the lines are needed in order to feed power to the unit. In this case, power is taken between phases 1 and 2.

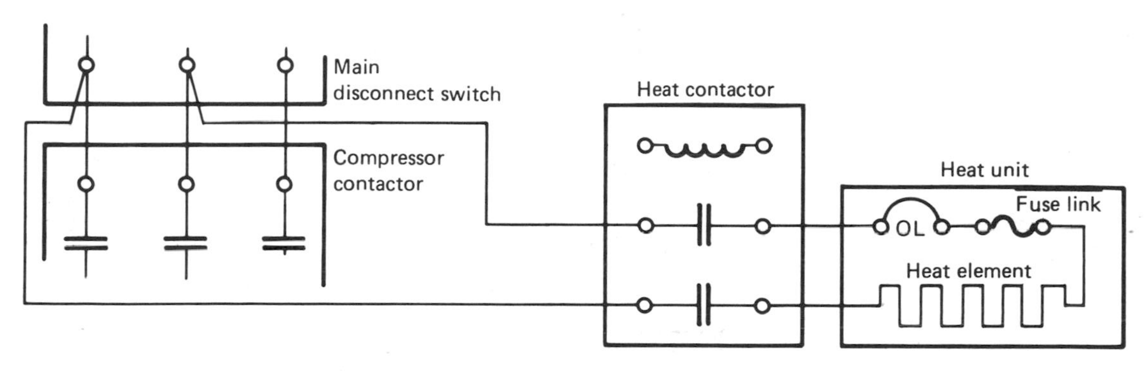

Figure 6–12. Power supplied to heat unit.

Heat strip 2 is connected in a circuit similar to heat strip 1. Heat strip 2 is connected between phases 2 and 3, as shown in Figure 6-13. Phase 2 is jumpered from a contactor terminal of heat strip 2 and the third phase is brought down from the main disconnect switch. The high-voltage heat connections are now complete.

LOW-VOLTAGE HEAT CIRCUITS

-Heat

The heat circuit is controlled by the thermostat. The *W*1 output of the thermostat controls heat strip 1, whereas the *W*2 output controls heat strip 2.

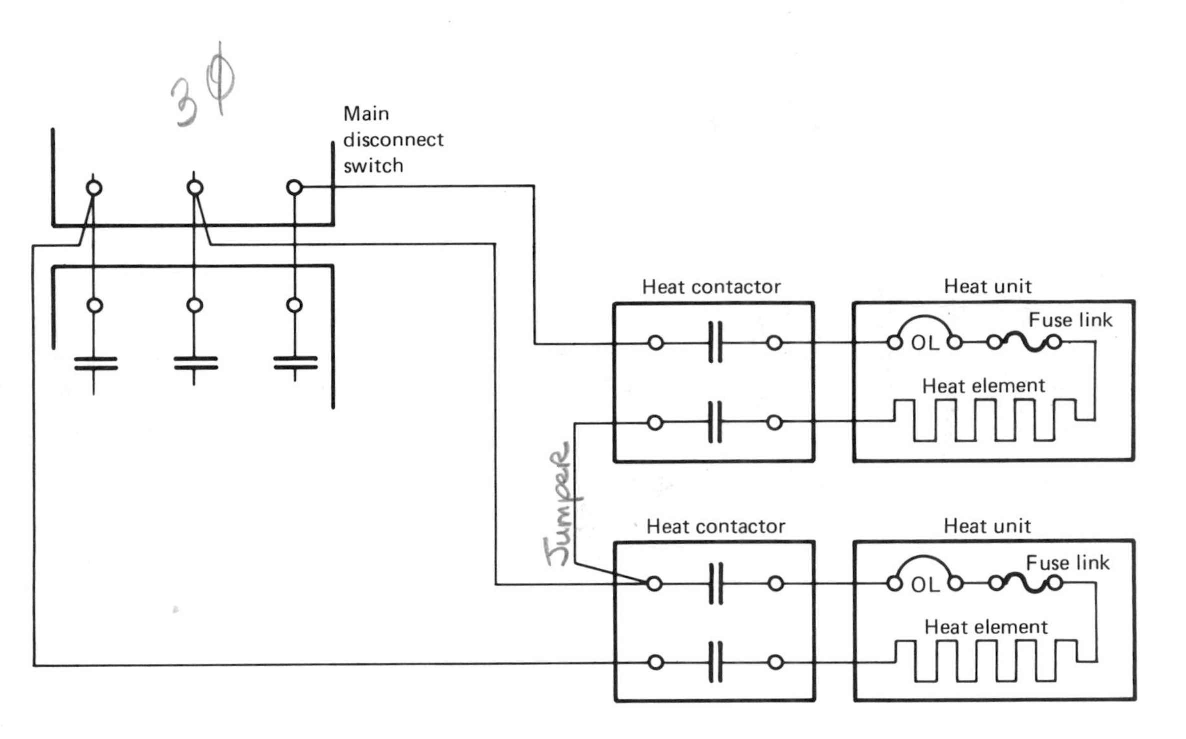

Figure 6–13. Connection of second heat unit.

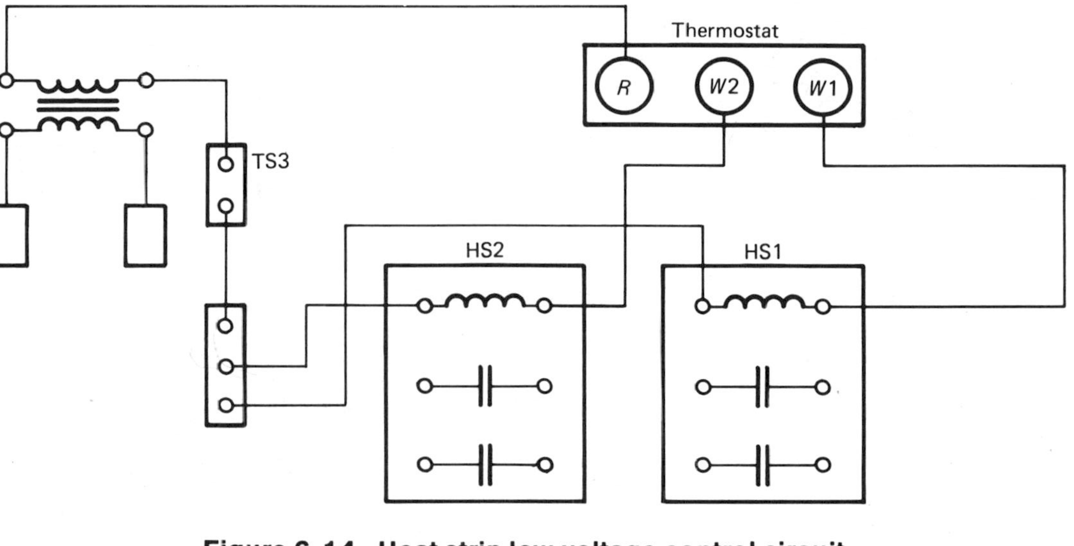

Figure 6–14. Heat strip low-voltage control circuit.

The control of the heat contactors by the thermostat is direct. The connection to the contactor coils is shown in Figure 6–14.

The single other control circuit for heat is associated with the inside blower motor. Since the same blower motor is used for cooling and heating, a heat relay is needed to interrupt the circuit to *G* on the thermostat. The heat relay ensures that there is no sneak circuit back through the thermostat that might turn the compressor on. The heat relay is connected as shown in Figure 6–15.

When "cool" is selected, the insider blower motor is controlled by the output at *G* on the thermostat. The circuit is through the normally closed contacts in the heat relay. When "heat" is selected and required, the output at *W*1 will cause the heat relay to energize. The normally closed contacts will open and the normally open contacts will close. Since the connection to *G* on the thermostat from the heat relay is now open, no sneak circuit can exist along this path.

The inside blower contactor will energize as a result of the control output at *W*1 through the now closed contacts of the heat relay. The inside blower motor will operate whenever there is an output of the thermostat at *W*1.

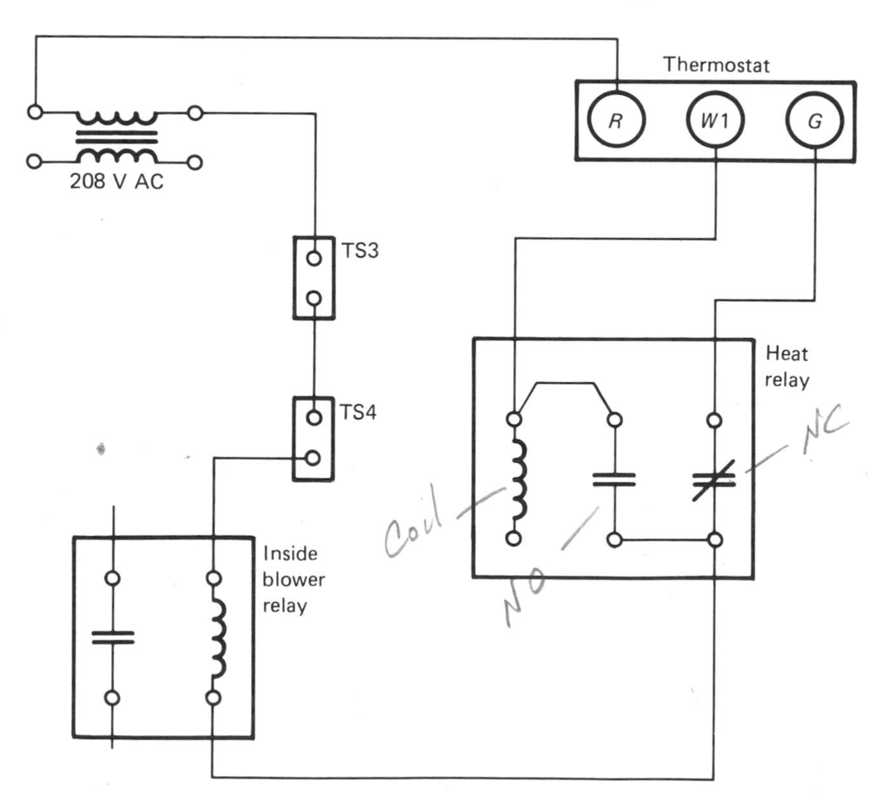

Figure 6–15. Addition of heat relay.

The cooling and heating system is now complete. Figure 6–16 is the overall wiring diagram. As a single diagram, it looks rather complex, but keep in mind that it is nothing more than the combination of circuits shown in Figures 6–1 through 6–15.

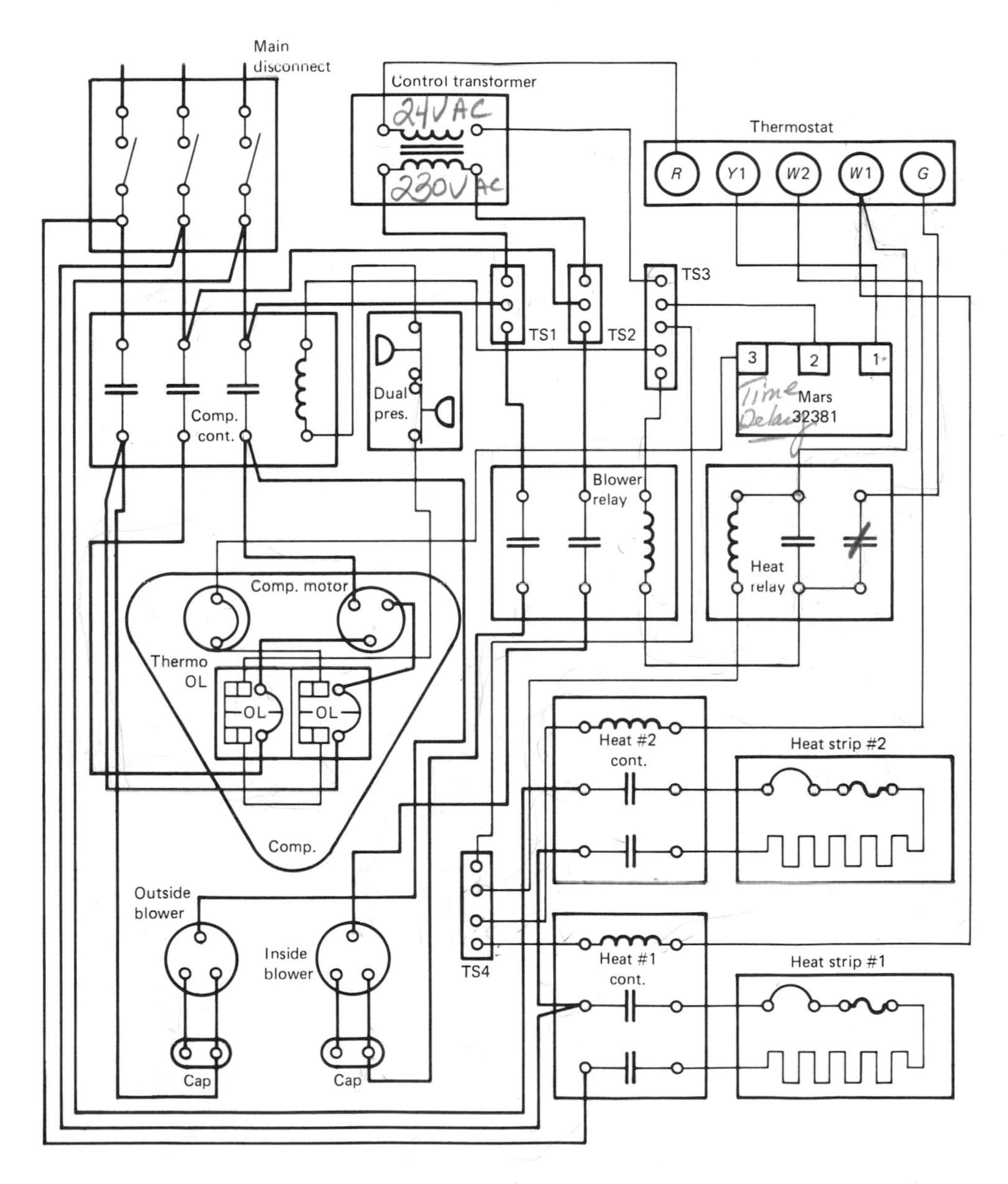

Figure 6–16. Three-phase single-stage cool, two-stage electric heat, air-conditioning system.

Chapter 7

GAS FURNACE CONTROLS

Gas furnace control has been developed to a level of safety that both the furnace manufacturing and service industries may be proud of. The standard controls associated with the flow of gas have established an excellent record of service. These controls include:

1. Pressure regulators
2. Gas valves (electric)
3. Thermocouples
4. Thermopiles (power thermocouples)
5. Thermopilot relays
6. Limit controls
7. Thermostats
8. High-voltage ignitor transformers

STANDARD SYSTEM

A common system includes a thermopilot valve, main solenoid gas valve, thermocouple, thermostat, and limit control. A sample system of gas furnace control is shown in Figure 7–1.

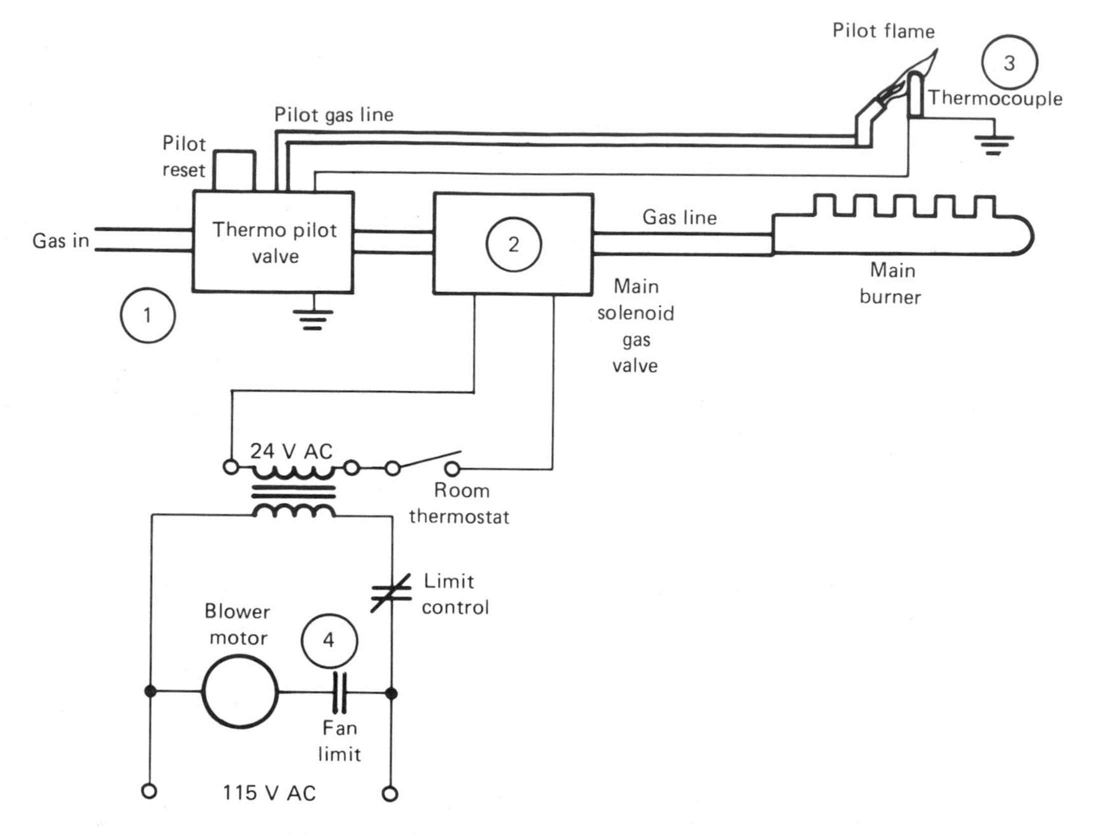

Figure 7–1. Gas furnace control system.

The thermopilot valve shown in Figure 7–2 controls the flow of gas to the pilot and to the main burner solenoid gas valve. The solenoid gas valve used in this system for main burner control is shown in Figure 7–3. Spring action keeps this valve closed when power is not supplied to the solenoid coil. When power is applied to the solenoid coil, the valve opens, allowing gas to flow as shown in Figure 7-3*b*).

The thermopilot valve, numbered 1 in Figure 7–1, controls gas flow to the pilot burner and to the main burner solenoid gas valve. The pilot may be lit when the reset button is pressed. Pressing the reset button allows gas to the pilot burner only.

The pilot burner, when lit, heats the thermocouple, numbered 3 in Figure 7–1. The heated thermocouple supplies thermoelectric current to the solenoid coil in the thermopilot valve. When the thermocouple is heated to a sufficiently high temperature by the pilot light, it will provide enough current to keep the solenoid coil in the thermopilot valve energized.

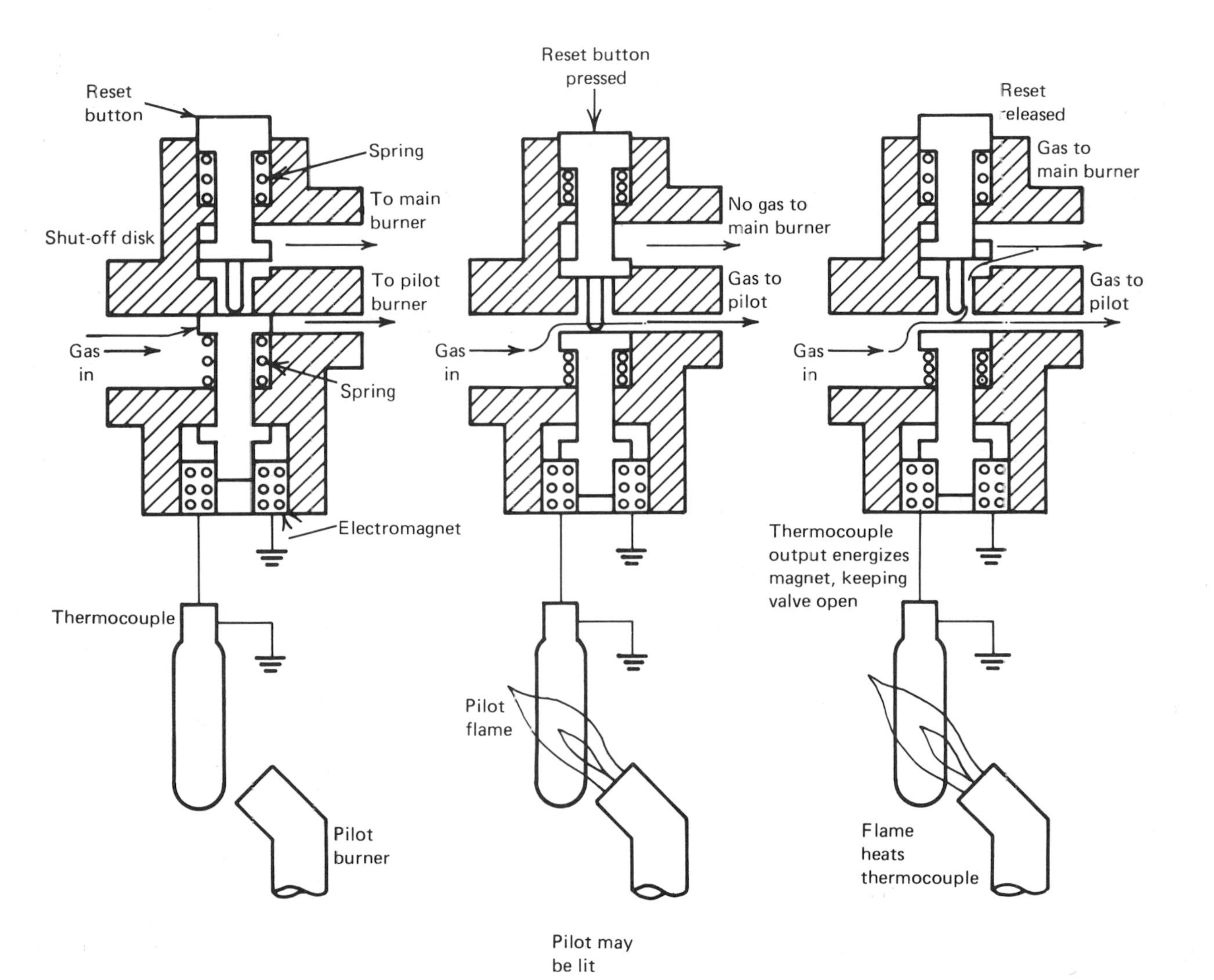

Figure 7–2. Action of thermopilot valve.

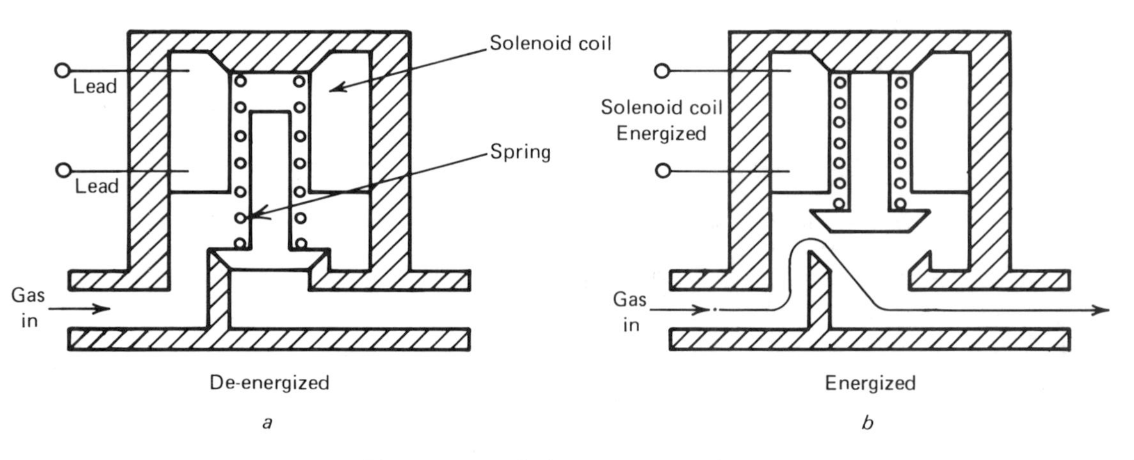

Figure7–3. Solenoid gas valve.

When this solenoid coil is energized, the release of the reset button allows gas to continue to flow to the pilot burner. Gas is also allowed to flow to the main burner solenoid gas valve, numbered 2 in Figure 7–1. The main burner solenoid gas valve is controlled by the thermostat. When the thermostat calls for heat, 24 volts AC is fed to the solenoid gas valve. The solenoid coil is activated, opening the valve and allowing the flow of gas to the main burner. Since the pilot is already lit, the main burner is ignited by the pilot.

The indoor blower operates after the plenum thermostat closes, 4 in Figure 7–1. This thermostat, often called a fan limit, closes when the air in the plenum is heated, thus keeping the blower from moving cold air into the space to be heated. A limit thermostat also senses plenum heat. If the temperature in the plenum becomes too high, the limit thermostat will open, removing primary voltage from the control transformer primary. Without primary voltage, there is no secondary control voltage. The main burner solenoid gas valve de-energizes and closes, cutting off the flow of gas to the main burner. The system is automatically reset when the plenum temperature is reduced to a level below the limit thermostat setting. Normally, the limit thermostat would operate only if the blower system should fail.

100 PERCENT SHUT-OFF SYSTEM

A further safety feature may be included in gas heating systems by providing 100 percent gas shutoff if a limit is exceeded. This system is shown in Figure 7–4. The limit control is connected in series with the thermostat solenoid

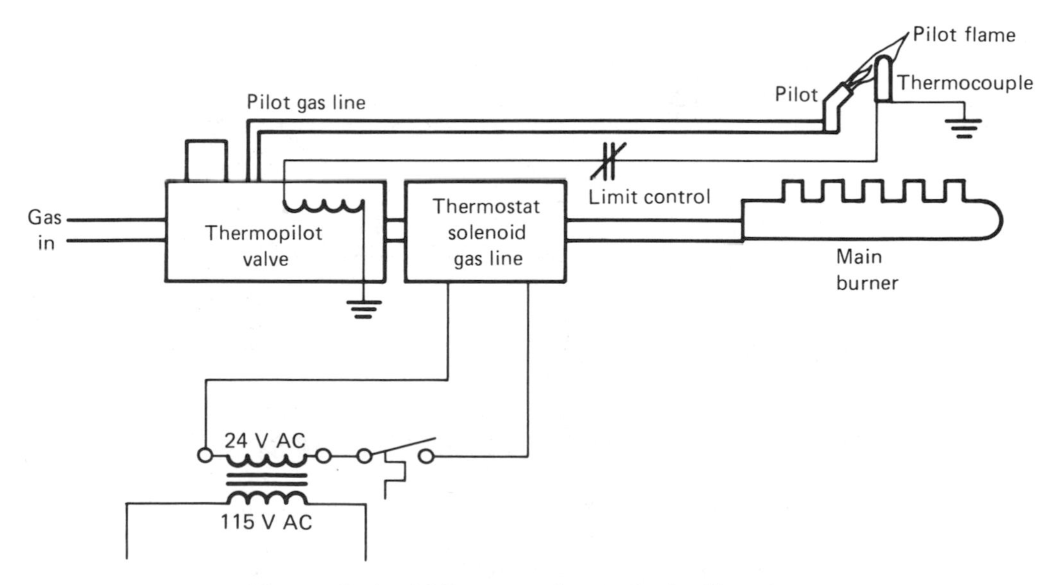

Figure 7–4. 100 percent gas shut-off system.

valve coil. The limit control, when open, will interrupt the circuit from the thermocouple. The thermopilot solenoid gas valve will de-energize and close. This shuts off gas to both the pilot and the main burner.

100 PERCENT SHUT-OFF, MULLIVOLT SYSTEM

Another safety feature is the two-valve millivolt system. In this system, both the thermopilot solenoid gas valve and main gas valve operate from the thermocouple current. The millivolt circuit is shown in Figure 7–5. If for any reason a limit is reached, the circuit to the thermopilot solenoid gas valve will be interrupted. When the thermopilot solenoid coil de-energizes, the valve will close, shutting off gas flow to both the pilot and the main burner gas lines. When the pilot goes off, the thermocouple will cool down. The decreased electric output from the thermocouple will cause the main burner solenoid coil to drop out.

The thermocouple used in this system may be a "power pile," which is nothing more than a series of thermocouple junctions connected in a series circuit. A single thermocouple will produce 30 millivolts across its output. Power piles may produce from 250 millivolts to 750 millivolts.

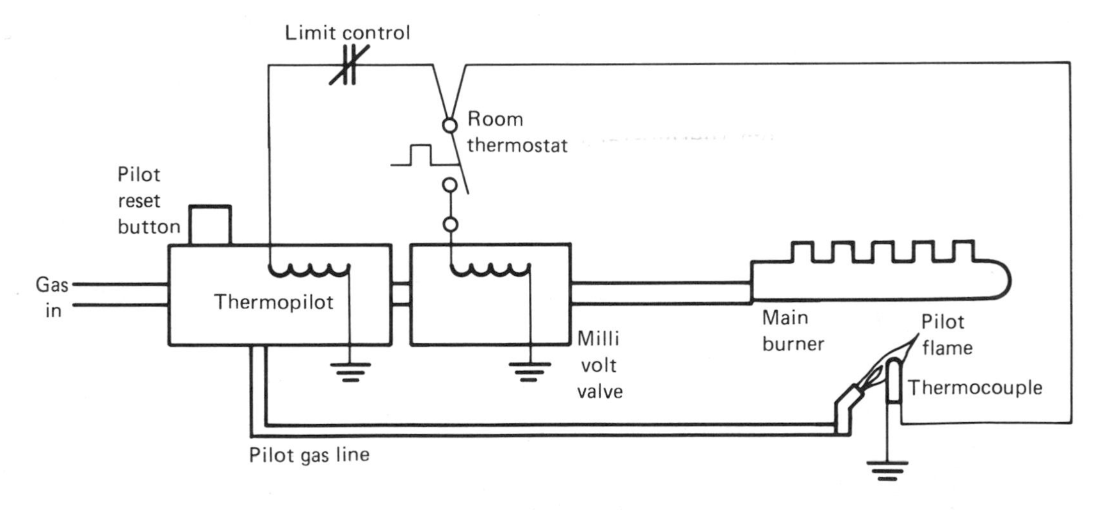

Figure 7–5. 100 percent shut-off system with dual valve control.

ENERGY-SAVING INTERMITTENT PILOT IGNITION SYSTEM

The new intermittent pilot gas systems are coming into general use as a means of conserving energy. The intermittent pilot gas system uses gas only when the system calls for heat. This system eliminates the continuously burning gas pilot light. The components that make up the system are:

1. Heat thermostat
2. Time delay
3. Buzzer
4. Plenum temperature limit
5. Capacitive discharge ignition
6. Ignition electrode
7. Thermocouple
8. Millivolt relay
9. Pilot gas valve with gas pressure switch
10. Main gas valve

NOTE

Figure 7–6 shows the general arrangement of components making up an energy-saving system. The colors given for color coded wires are used as a reference in this drawing only. They are not to be considered standard for energy-saving systems. The drawing is not meant to be a copy of any manufacturer's system. The individual manufacturer's service bulletin should be referred to when servicing equipment.

The control transformer (1) in Figure 7–6 provides 24 volts AC for operation of the energy-saving system. The red wire output from the transformer feeds into the room thermostat (2). When the thermostat calls for heat, the 24 volts is fed out of the thermostat on the white wire to the time delay (3). The circuit continues through the normally closed contacts of the time delay to the capacitive discharge ignition unit (4). The return side (black wire) of the capacitive discharge ignition unit is connected through the normally closed contacts 2 and 1 of the millivolt relay (5) to the control transformer. When power is fed to the capacitive discharge ignition unit, it produces a high-voltage output. The high voltage is fed to the electrodes (11), which emit sparks to ground.

The 24 volts is at the same time fed from the input of the capacitive discharge ignition unit on the white/red wire to the main gas solenoid valve (9) and the pilot gas solenoid valve (8). The return side of the pilot gas solenoid valve is permanently connected to the return side of the 24-volt control transformer with a black wire. The pilot gas solenoid valve will energize, providing gas to the pilot and to the input of the main gas solenoid valve. A pressure switch (12) in the pilot gas valve will close if the gas pressure is high enough. The pilot will light with gas from the pilot gas solenoid valve and spark from the capacitive discharge ignition system. The flame from the pilot light will heat the thermocouple (7). The low-voltage output of the thermocouple is fed through the gas pressure switch (12) to the millivolt relay. When the thermocouple is sufficiently hot, the millivolt relay will energize, breaking the 1 and 2 contacts and making the 1 and 3 contacts. The capacitive discharge ignition will stop producing high voltage and the sparking at the electrodes will stop. With the 1 and 3 terminals of the millivolt relay making contact, the main solenoid gas valve has a return connection to the 24-volt control transformer. The main gas solenoid valve opens, and the main burner lights because the pilot flame is producing the required heat. When the building is heated, the room thermostat contacts will open. This removes the 24 volts from the system. Both the pilot gas valve and the main gas valve will close, shutting off all gas flow.

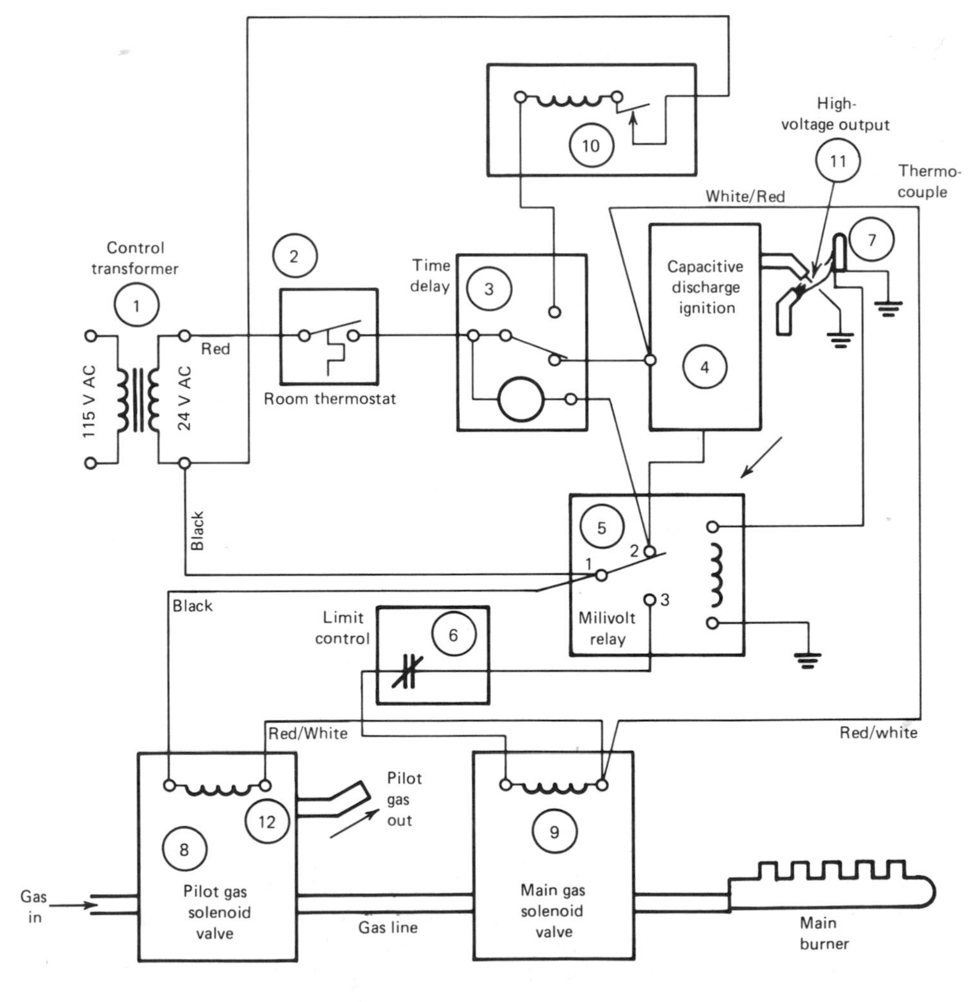

Figure 7–6. Energy-saving system.

SAFETY FEATURES

The plenum temperature limit (6) provides the same protection as it does when used with the standard gas heat system. If, for any reason, the plenum temperature should rise above the preset limit, the contacts of the limit control would open the system. The main gas valves would close, shutting off the flow of gas to the main burner.

The gas pressure switch (12) is located in the housing of the pilot gas valve. If gas pressure should drop, the gas pressure switch would open. The millivolt relay would de-energize, removing the connection from 1 to 3 of the relay. This would interrupt the return line connection from the main gas valve and the valve would close, shutting off the flow of gas to the main burner.

The electrodes of the high-voltage ignition system would immediately start to spark at the ignition electrode in order to restart or ensure continuous flow at the pilot light. When the gass pressure returned to normal, the pilot would light, heating the thermocouple, or if the pilot had remained lit, the millivolt relay would energize when the gas pressure switch reclosed. The high-voltage ignition would again cut off, and the main gas valve would again open.

If, for any reason, the millvolt relay does not energize after the room thermostat closes, the time delay will open the electric circuit to the system after 90 seconds. If the time-delay motor (3) is allowed to run for 90 seconds, a buzzer (10) will sound, indicating a problem. Under normal conditions, the time-delay motor only runs for a few seconds as the thermocouple heats, that is, until the thermocouple provides sufficient voltage to energize the millvolt relay. The return line connection for the time-delay motor is through the 2 and 1 contacts of the millvolt relay. This circuit is opened shortly after the pilot is lit. The time-delay motor is springloaded to return to zero.

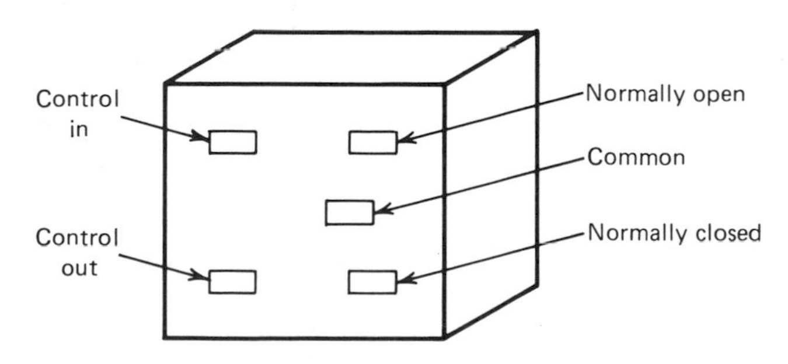

Figure 7-7. Electronic time delay with internal spot contacts.

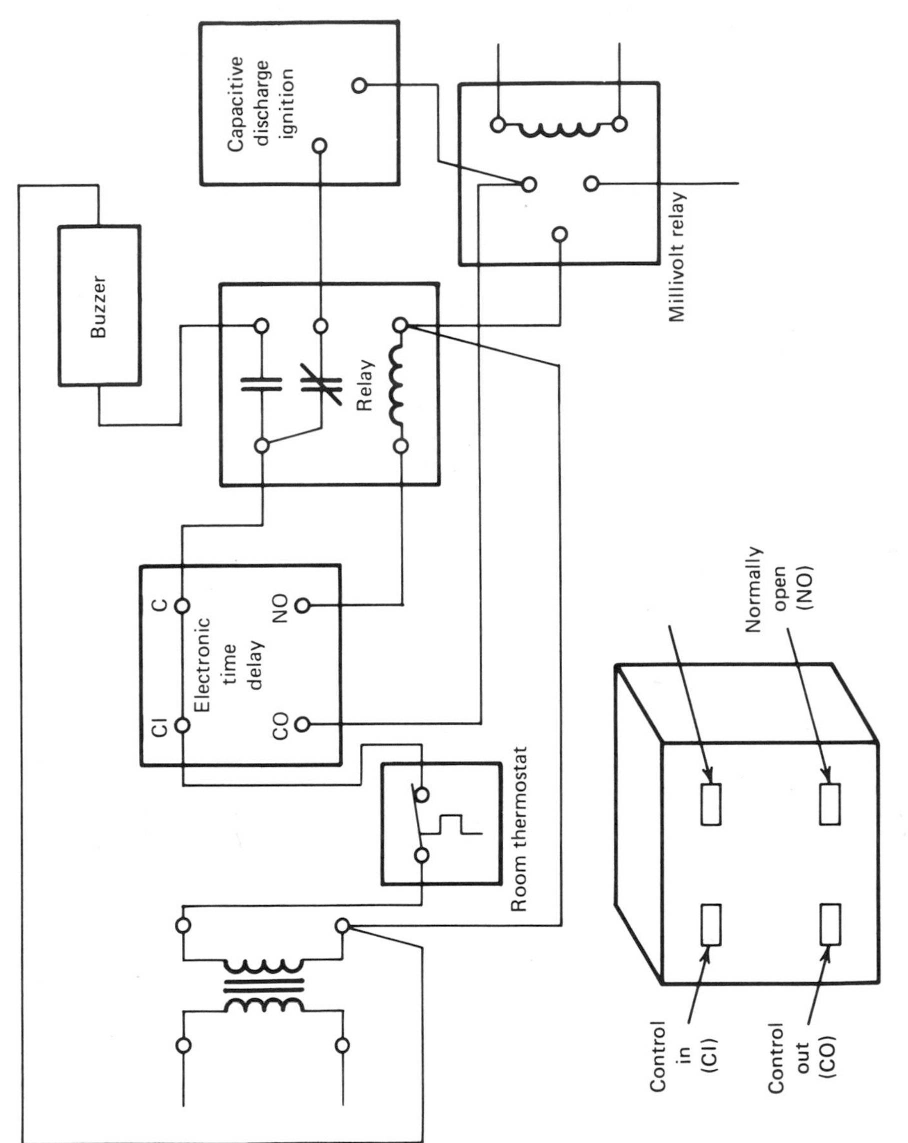

Figure 7–8. Electronic time delay circuit with external relay.

MODIFIED SYSTEMS

Some of the latest energy-saving systems use electronic time-delay relays. The device is not repairable and must be replaced if it malfunctions. The connections to the electronic time delay are shown in Figure 7–7. The unit can be used as a direct replacement for the mechanical time delay shown in Figure 7–6.

An electronic time delay with a single set of normally open contacts could also be used, along with an external relay in the circuit. The electronic time delay would only control the external relay, which would perform the transfer function in the circuit. An example is shown in Figure 7–8.

Two recent developments are the hydraulic flame sensor and the electronic flame sensor.

HYDRAULIC FLAME SENSOR

The hydraulic flame sensor uses a bulb and capillary assembly to control an electric switch. When the pilot flame heats the bulb, the fluid in the bulb

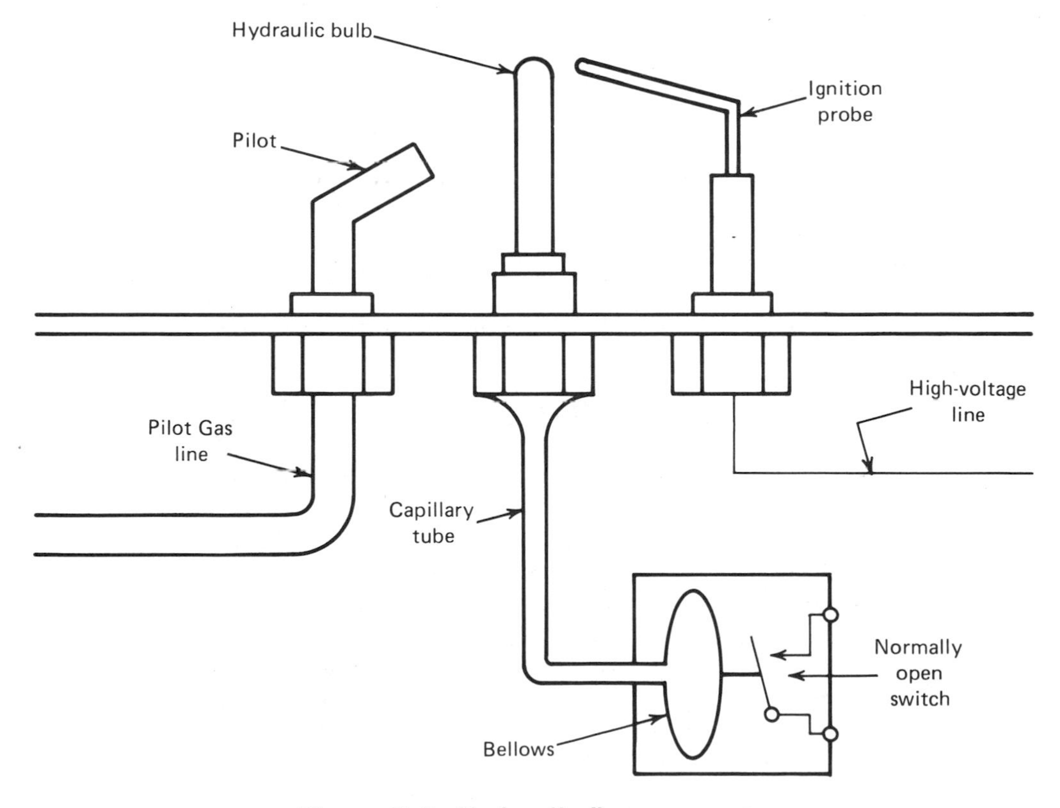

Figure 7–9. Hydraulic flame sensor.

expands, increasing pressure. This pressure controls a bellows-actuated switch. The switch in turn controls the application of power to a relay coil. The hydraulic flame sensor is shown in Figure 7–9. A part of the control circuit that includes the hydraulic flame sensor is shown in Figure 7–10. The switch controls the application of power to the relay, and the relay cuts off the capacitive discharge system. The relay also completes the circuit that opens the main burner valve.

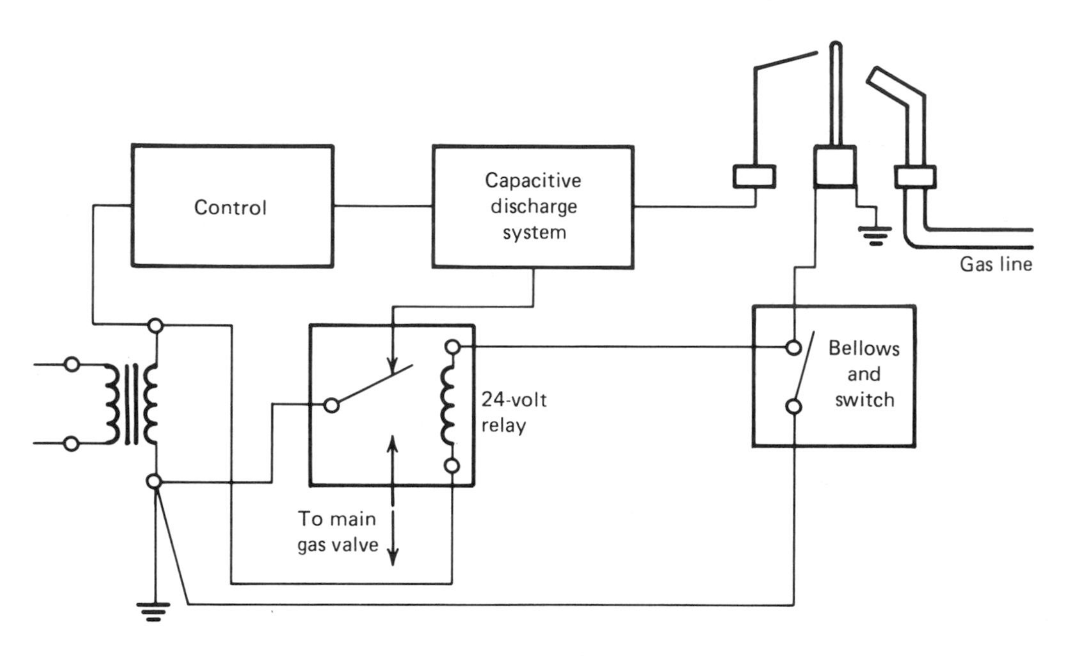

Figure 7–10. Hydraulic flame detector circuit.

ELECTRONIC FLAME SENSOR

The purpose of a flame sensor is to prove that the pilot is lit. The electronic flame sensor does this by using the pilot flame as the medium of electrical conduction. An electronic flame sensor is shown in Figure 7–11.

It is a well-known fact that a flame is a chemical reaction. In the chemical reaction of a gas pilot flame, positive electrical ions are developed. In a circuit, the positive ions are attracted to the conducting surface with the largest surface area. The movement of the positive ions to the conducting surface with the largest area provides a conduction path in one direction. The process is sometimes referred to as flame rectification. In the gas pilot system, the metal pilot burner is the metal surface with the largest area. The positive-ion movement is from the smaller probe to the pilot burner. The pilot burner is grounded. The circuit action is shown in Figure 7–12.

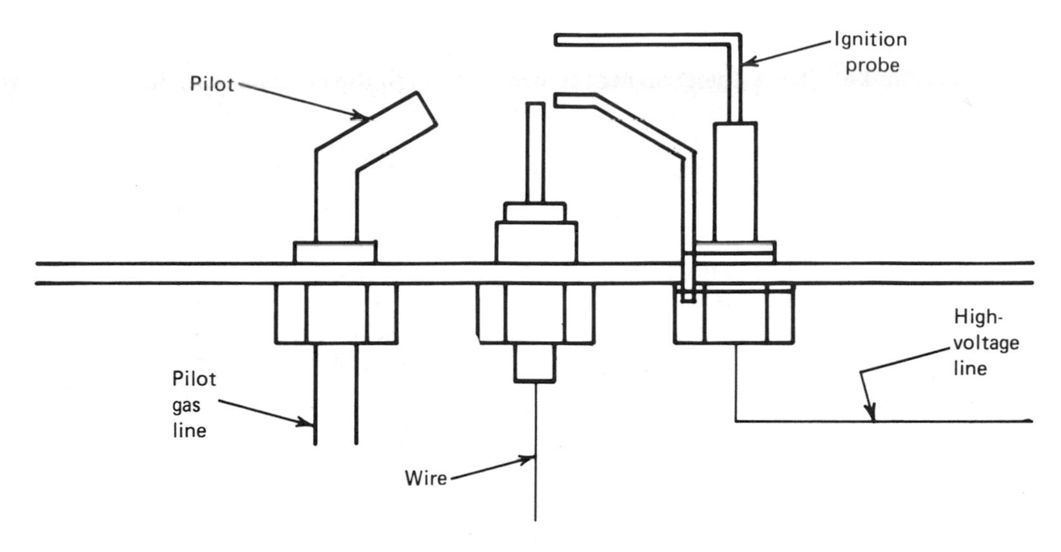

Figure 7–11. Electronic flame sensor.

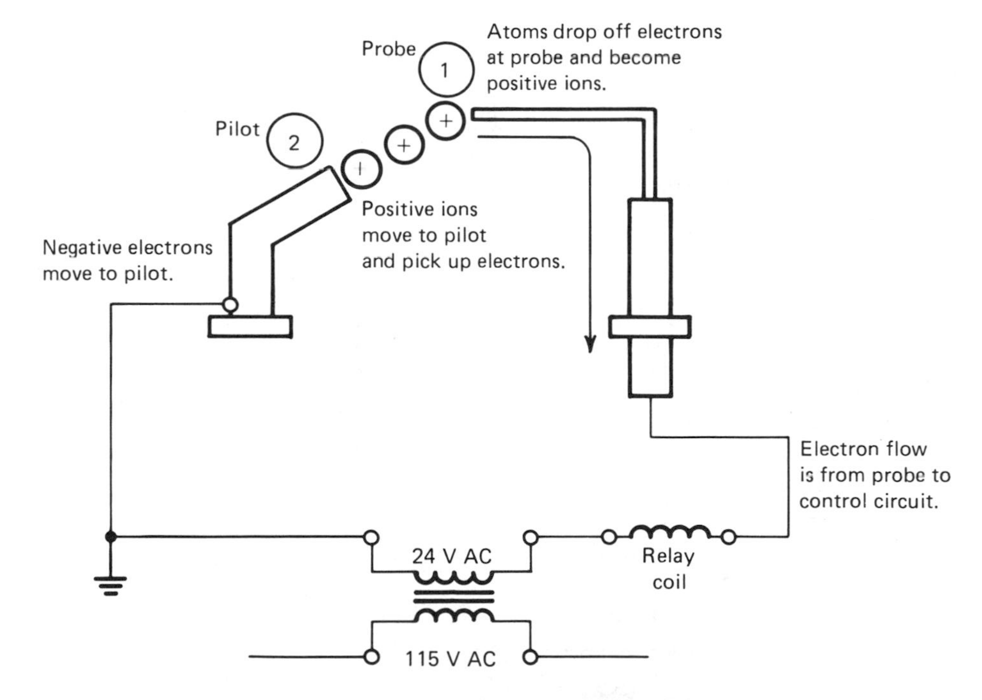

Figure 7–12. Flame rectification.

The positive ions continuously form in the flame by dropping off elec trons at the probe (1). The positive ions move to the metal housing of the pilot light (2) where they pick up electrons, thereby becoming a neutral atom again. The process continues with electrons moving from ground to the pilot, and electrons then leaving the probe through the external circuit.

In the external circuit, the electron current is fed from the probe through the relay coil and through the low-voltage transformer secondary, as shown in Figure 7–13. The relay in Figure 7–13 provides the same function as the millivolt relay shown in Figures 7–6 and 7–8. The purpose is to control the shutdown of the capacitive discharge ignition and to complete the control circuit, thus opening the main burner solenoid gas valve.

A safety feature added to this circuit is the diode placed across the relay coil. The diode is placed in the circuit because the normal electron flow resulting from flame rectification cannot pass through it. This situation is

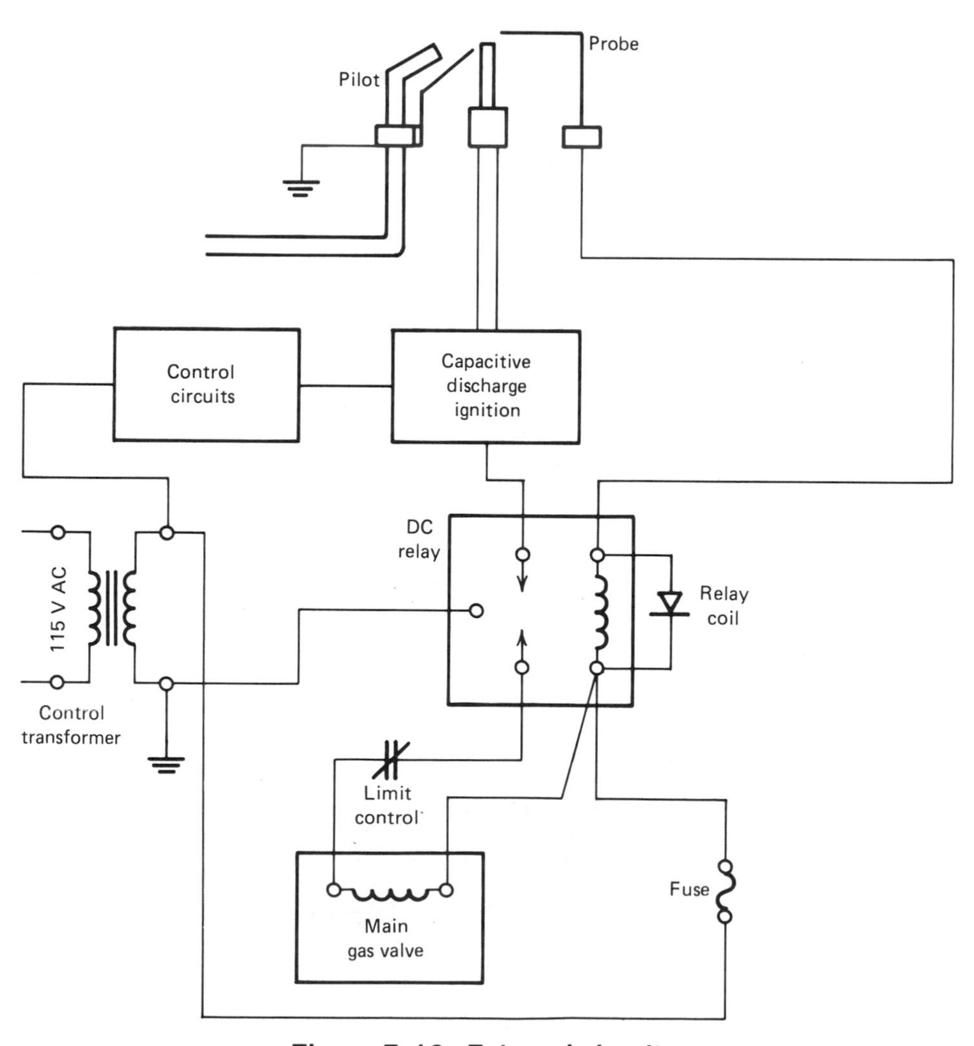

Figure 7–13. External circuit.

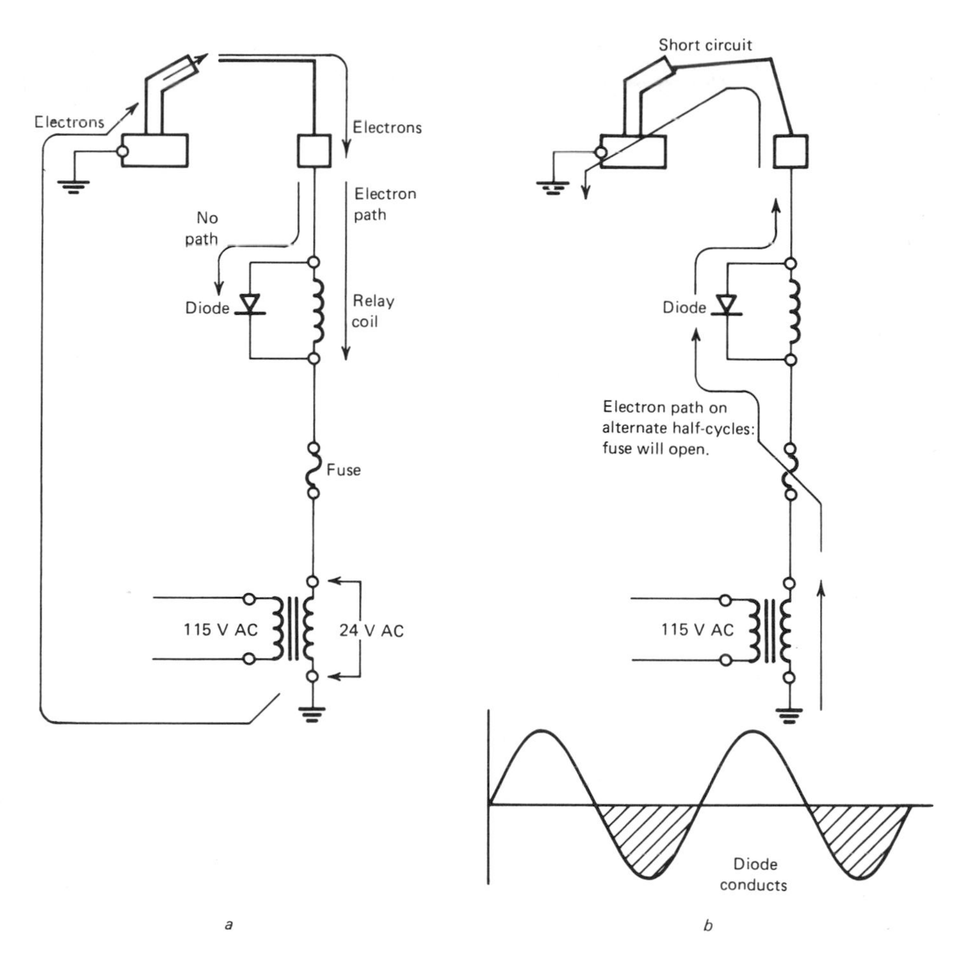

Figure 7–14. Addition of diode to relay coil.

shown in Figure 7–14*a*. Remember that electrons can only flow through a diode in the direction toward the arrow head. The electron flow resulting from the flame rectification process cannot go through the diode; it must go through the relay coil. If the probe should be bent, making contact with ground (short circuit), there would be a current path during both alternations of the AC input. On every other alternation of the input, there would be a short path through the diode as shown in Figue 7–14*b*. Sufficiently high current would flow, causing the fuse to open. All power would then be removed from the system, and it would be shut down.

New controls are rapidly being developed for the heating industry. Until such a time when the controls become standardized, it is necessary that the technician always refer to the manufacturer's information sheet concerning the particular unit being installed or serviced.

Review Questions

Chapter 7

Mark each statement T for true or F for false.

1. The thermocouple produces a gas output that controls the pilot flame. 1. __
2. The thermopilot gas valve feeds gas directly to the main burner. 2. __
3. The main burner gas valve is controlled by either the single thermocouple or a thermopile in the millvolt system. 3. __
4. The high-voltage capacitive discharge ignition operates whenever its main burner is turned on. 4. __
5. A plenum heat limit thermostat is not needed in a capacitive discharge ignition gas system. 5. __
6. All electronic time delays must be modified for use with gas heating systems. 6. __

Chapter 8

CONDITIONED AIR DELIVERY DIAGRAMS

A very important part of an air-conditioning technician's responsibility is the delivery of conditioned air to the space to be heated or cooled. To a great extent, the overall efficiency of the system is dependent on how well an air delivery system operates. It is of little benefit to have an evaporator system operating at maximum efficiency if the cooled air does not reach the space to be air conditioned; more importantly, there is no reason to have the evaporator system operating at all if the outside air temperature is low enough to cool the space to be air conditioned. The air delivery system is the final phase of the complete heating and cooling system.

Figure 8–1 is the diagram of a complete conditioned air delivery system for an office or apartment building. The overall unit could be located in a penthouse or in a basement. The system would be approximately the same for both.

AIR DELIVERY SYSTEM COMPONENTS

The following paragraphs briefly describe the components of the air delivery system.

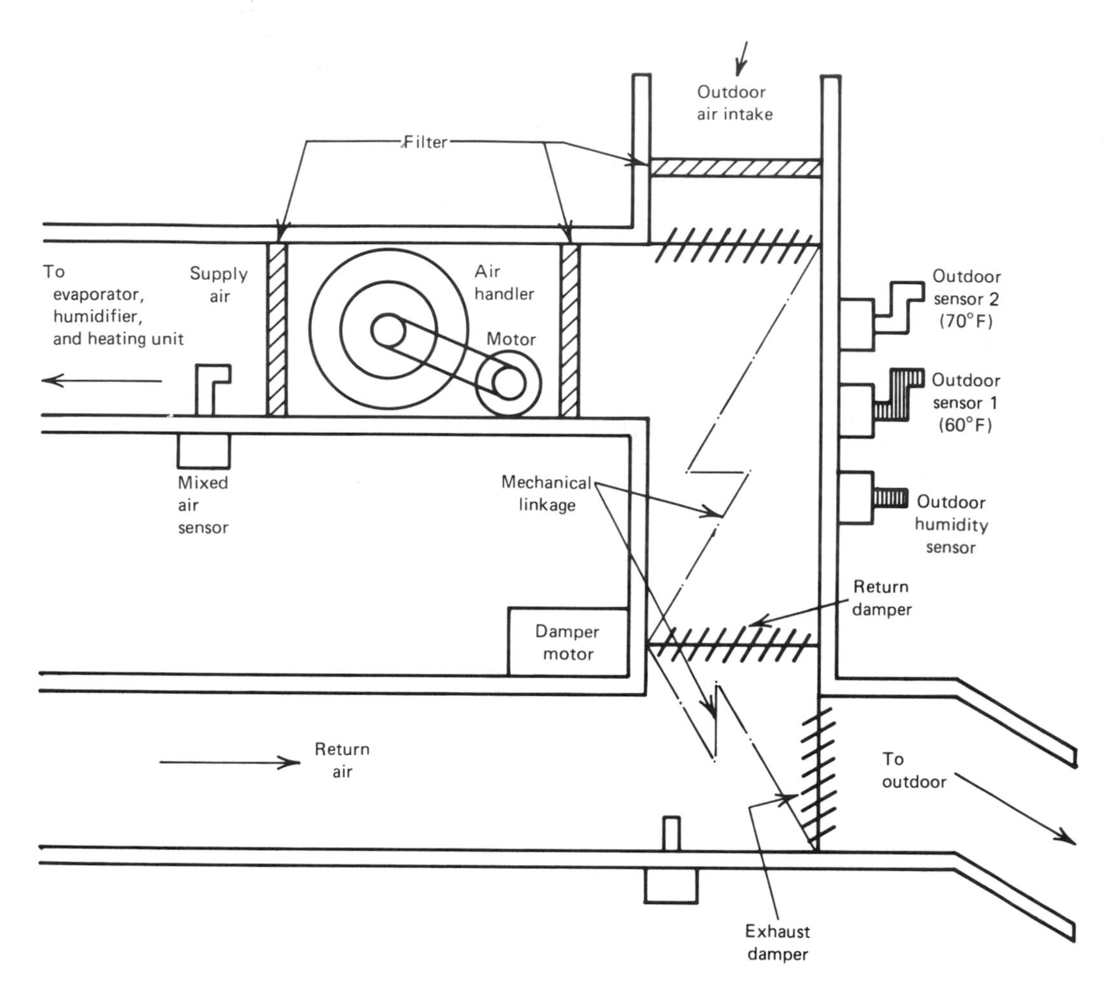

Figure 8–1. Conditioned air delivery system.

Air Handler:

The air handler consists of the motor and fan system and associated controls. The air handler must provide the volume and air speed required by the system size.

Mixed Air Sensor:

The mixed air sensor is a control whose operation is based on the temperature of the air being supplied to the conditioner. The mixed air is a combination of return air and outside air. This device, a thermostat, is usually set for about 60°F.

Outdoor Ambient Temperature Sensor 1

Outdoor ambient temperature sensor 1 is a control whose operation is based on whether the "free" air is cold enough to be used for cooling. This device is usually set for about 60°F.

Outdoor Ambient Temperature Sensor 2:

Outdoor ambient temperature sensor 2 limits the amount of outside air allowed into the system. When the temperature of the outdoor air reaches the preset limit, the outdoor dampers close to the minimum position, reducing the amount of warm outdoor air allowed into the system. This device is usually set for about 70°F.

Humidity Sensor:

The humidity sensor controls the humidifier. The humidity sensor determines the amount of moisture in the return air. If the moisture level is below a preset limit, the humidifier will be turned on.

Outdoor Humidity Sensor:

The outdoor humidity sensor senses the condition of the outdoor air and determines whether the compressor will be turned on when outdoor air is used for cooling.

Humidifier:

The humidifier ensures that water will be introduced into the air flow as required. The humidifier is used for heating.

Filters:

The filters are used to keep foreign particles that might clog up the heating and cooling elements out of the system.

Exhaust Damper:

The exhaust damper controls the amount of return air that is discharged to the outdoor through a duct, as shown in Figure 8–2.

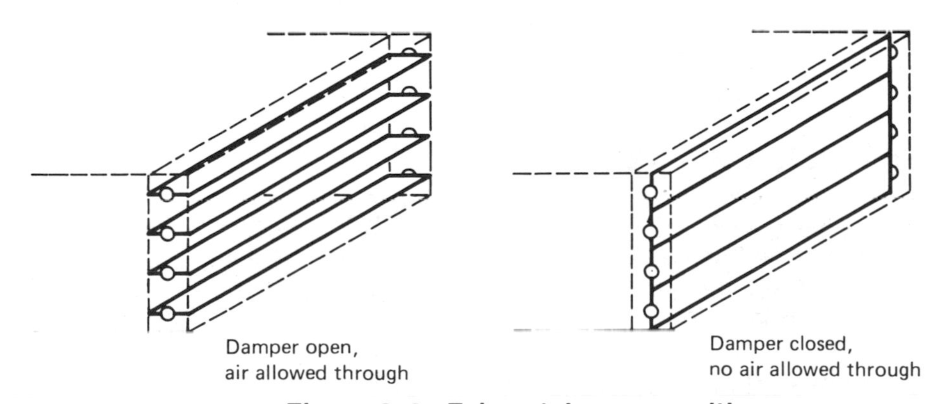

Figure 8–2. Exhaust damper positions.

Return Damper:

The return damper controls the amount of air from the conditioned area that is going to be recirculated. The return damper is mechanically coupled with the exhaust damper. When the return damper closes, the exhaust damper opens.

Outdoor Damper:

The outdoor damper controls the amount of outdoor air that is allowed to enter the system. A fixed amount of fresh air is always introduced into the system. The outdoor damper is mechanically linked to the return damper. When the return damper closes, the outdoor damper opens. When the system is "off," the outdoor damper closes.

Damper Motor:

The damper motor is designed specifically for damper system use. The standard motor has three wire inputs. The motor circuit is shown in Figure 8–3. When power (24 volts) is applied between terminals 1 and 2, the motor will

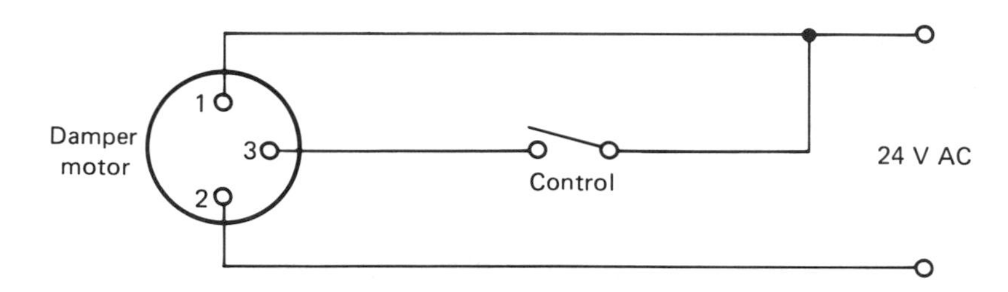

Figure 8–3. Damper motor circuit.

rotate to a limit in approximately 45 seconds. When terminal 3 is connected through the control circuit, the motor will rotate to the opposite limit in 45 seconds. When terminals 1 and 2 are active alone, the outdoor damper and exhaust damper will close and the return damper will be fully open. When terminal 3 is connected through the control circuit, the return damper will close and the outdoor and exhaust dampers will open.

MODULATING DAMPER ACTION

A modulating damper is in almost continuous motion, opening and closing in order to maintain the required mixed air temperature. The mixed air sensor controls the damper action as the air temperature at the sensor changes with changes in the damper position. The outdoor air damper, return air damper and exhaust air damper will often be in continuous motion when the system is in the cooling mode of operation. The dampers actions are controlled by

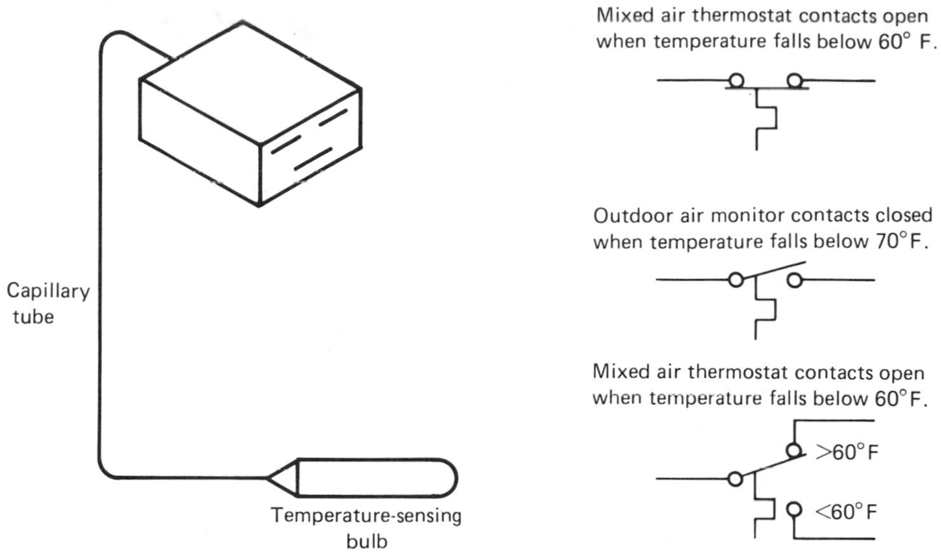

Figure 8–4. Fixed temperature controls—factory set.

temperature measurements made with fixed temperature controls. Fixed temperature controls are shown in Figure 8–4.

CONTROL CIRCUITS

The system of air delivery shown in Figure 8–1 is an example of an energy-saving system. Whenever the outside air is cool enough, it will be used to cool or help cool the inside of the building. The operation of the system is controlled by temperature and humidity sensors located inside the air ducts and by sensors that respond to outdoor air temperature and humidity.

The complete circuit for this damper motor system is shown in Figure 8–5. As with other complete circuit diagrams, for the energy-saving system circuit diagram appears complex. The circuit is easily understood when the individual units of operation are investigated.

DAMPER MOTOR CONTROL

There are four basic weather conditions during which the damper control system is important:

1. Outdoor air temperature below 60°F with low humidity.
2. Outdoor air temperature below 60°F with high humidity.

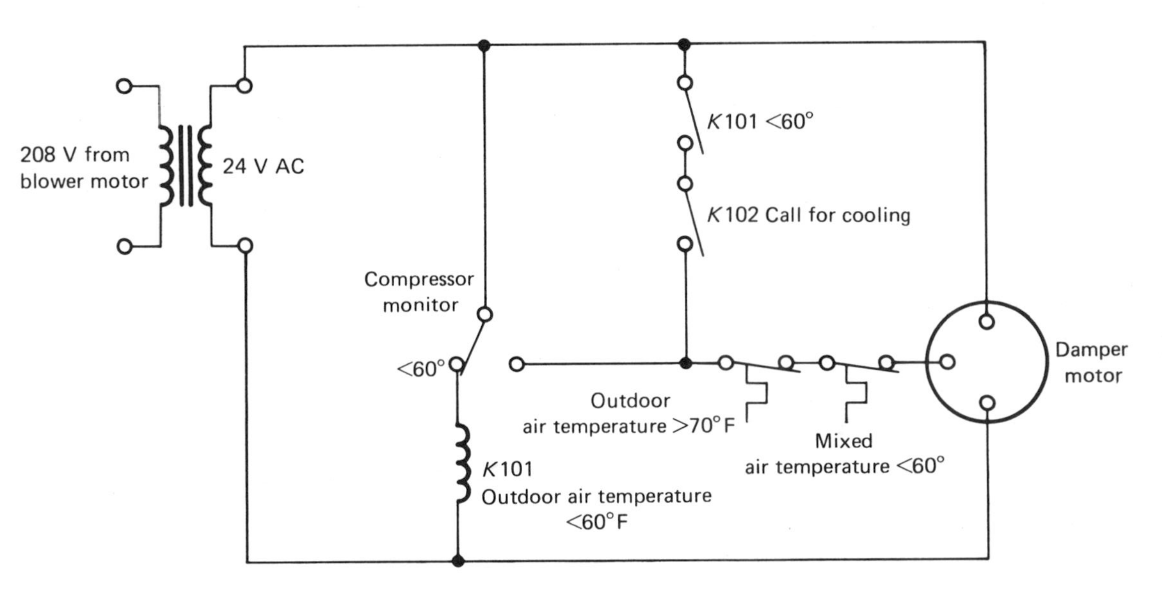

Figure 8–5. Damper motor control.

3. Outdoor air temperature between 60°F and 70°F.
4. Outdoor air temperature above 70°F.

Outdoor Air Temperature Below 60°F with Low Humidity

The power source for the damper system is the step-down transformer that receives its primary voltage from the same source as the indoor blower motor. The connection of the transformer primary is shown in Figure 8–6. Whenever the indoor blower relay is energized, the damper low-voltage transformer primary will be connected across the 208-volt line. When an energy-saving damper system is used, the indoor blower motor operates continuously whenever the main disconnect switch is closed. This is equivalent to having the thermostat fan switch in the "on" position. The 24 volts AC is then available at the secondary to operate the damper motor. With the 24 volts AC connected to terminals 1 and 2 of the damper motor, the motor rotates to the "start" limit. In this position, the return damper is fully open. The exhaust damper is closed, and the outdoor damper is opened the minimum amount, providing a limited amount of fresh air to the system.

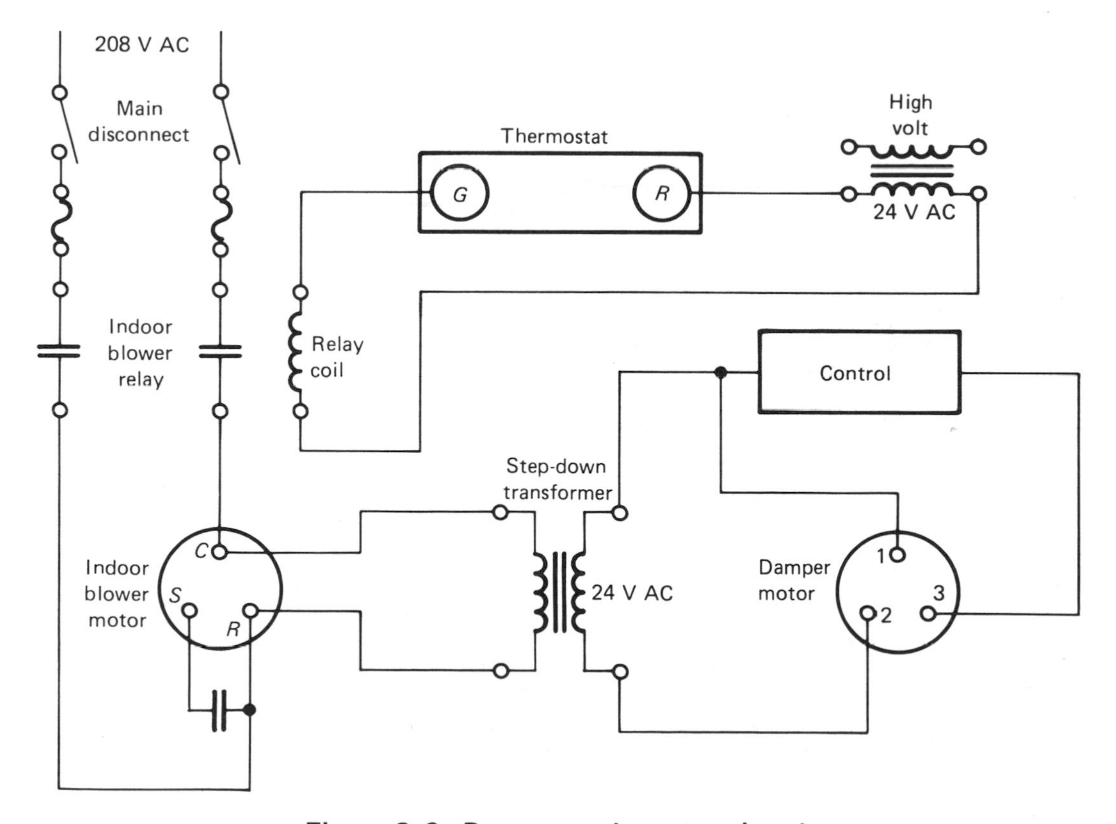

Figure 8–6. Damper motor power input.

An outdoor air temperature below 60°F is low enough to meet the cooling needs of the building without the operation of the compressor. The damper control system must keep the compressor from operating and must open the outdoor damper, close the return damper, and open the exhaust damper.

The control of the compressor is provided by relay *K*103 (see Figure 8–7). The normally open contacts 1 and 3 of relay *K*103 will be open if the outside air temperature is below 60°F. These contacts are in series with the *Y*1 output of the thermostat. A complete circuit will not exist to the compressor contactor coil. With the contacts of *K*103 open, the compressor contactor cannot be energized. The contacts of relay *K*103 are at number 1 in Figure 8–7.

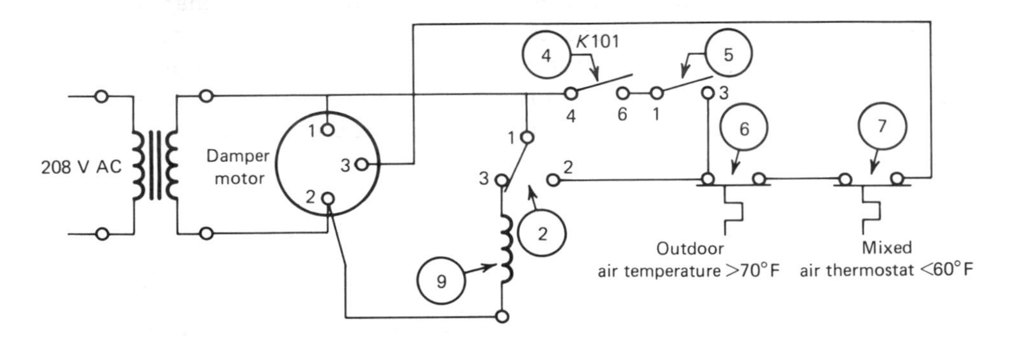

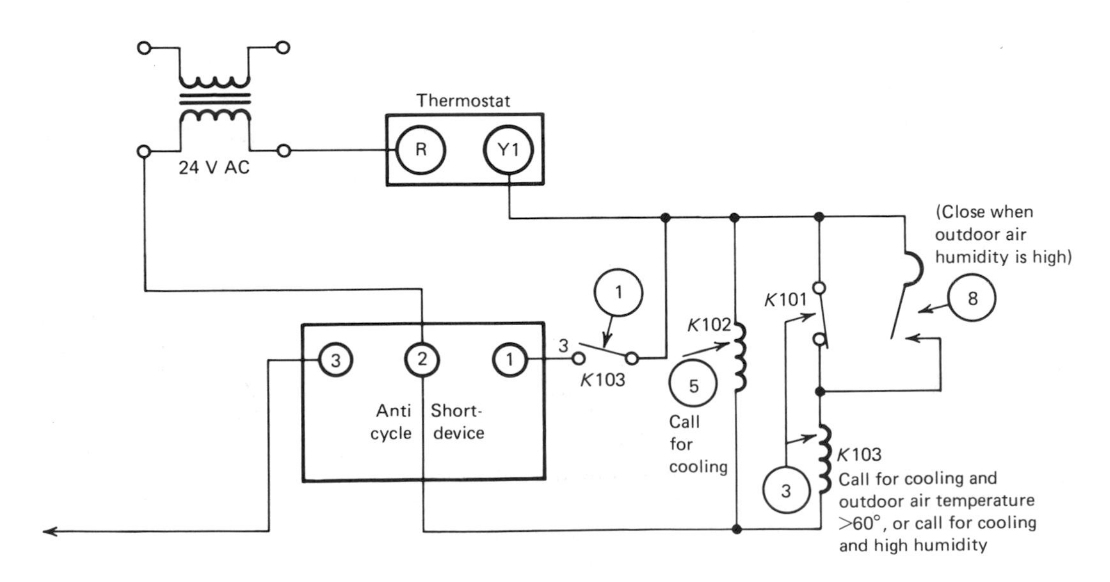

Figure 8–7. Compressor and damper control.

Relay *K*103 is controlled by the outdoor air thermostat, numbered 2 in Figure 8–7. When the outdoor temperature is below 60°F, the 1 and 3 contacts of the thermostat switch will be closed, providing 24 volts AC across relay *K*101. Relay *K*101 will energize and open the normally closed contacts 1 and 2, which are in series with the coil of relay *K*103. This is shown as 3 in Figure 8–7. Relay *K*103 will stay de-energized, since relay *K*101 is energized and the outdoor air has low humidity.

The resulting situation is that the thermostat has a call for a cooling condition. The compressor will not operate the relay, since *K*103 contacts are open. The thermostat output at *G* calls for indoor blower operation, which provides a 208-volt input to the damper supply transformer (see Figure 8–6). The damper motor is in the normal position, the return damper is open, and both exhaust damper and outdoor damper are closed. In order to obtain cooling from outside air, the outdoor damper must be opened, the return damper closed, and the exhaust damper opened. This can be accomplished by providing 24 volts AC to terminal 3 of the damper motor. Relay *K*101 shown at 2 in Figure 8–7 is energized since the outdoor air temperature is below 60°F. A set of normally open contacts 4 and 6 of relay *K*101 is shown (at 4 in Figure 8–7) in series with the normally open contacts 1 and 3 of relay *K*102 (at 5 in Figure 8–7). Relay *K*102 is energized whenever there is a call for cooling.

The circuit to terminal 3 of the damper motor continues through the contacts of the thermostat outdoor air temperature greater than 70°F and the contacts of the mixed air thermostat. This part of the damper circuit is shown in Figure 8–8. The damper motor will rotate and open the outdoor damper. The return damper will close and the exhaust damper will open.

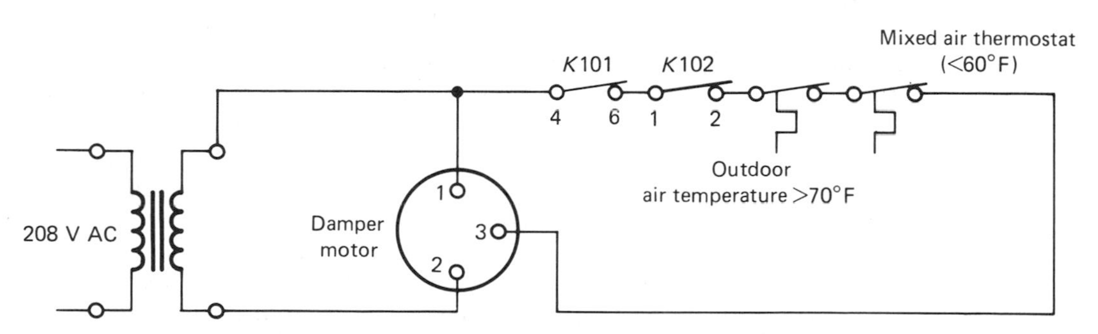

Figure 8–8. Damper connections for use of outdoor air.

When the outdoor damper is opened, cool outdoor air (temperature below 65°F) is introduced into the cooling system. The return damper is closed, blocking the warmer indoor air from recirculating, and the exhaust damper opens to release the warm indoor air to the outdoors. This system is shown in Figure 8–9.

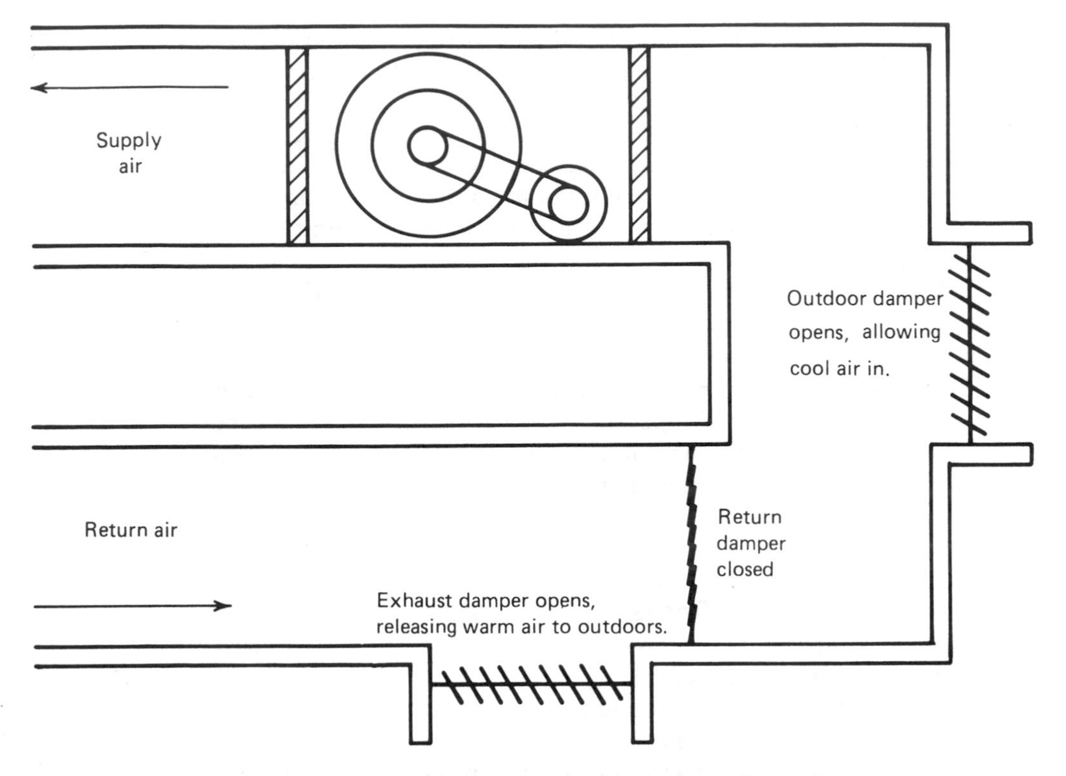

Figure 8–9. Damper positions for use of outdoor air.

After sufficient outdoor air has entered the system, the cooling effect will reduce the mixed air temperature to below 60°F. The thermostat will open, and the circuit to terminal 3 of the damper motor will be incomplete. Since power is still being applied to terminals 1 and 2 of the damper motor, the motor will rotate to close the outdoor damper. At the same time, the return dampers will be opening, and the exhaust dampers will be closing. Warmer return air will be mixed with outdoor air. The outdoor damper will continue to close, and the return damper will continue to open until the air reaching the mixed air thermostat (as shown in Figure 8–8) reaches a temperature above 60°F. The thermostat will then close again. The damper will again have a power connection at terminal 3 and will start to rotate in a direction to open the outdoor damper, allowing in more cool outdoor air.

The damper system will modulate, opening and closing, thus keeping the temperature at the mixed air thermostat varying around 60°F. The inside of the building will be cooled solely by the cooler outdoor air. The compressor will not operate and energy will be saved.

The use of outdoor air for cooling when the humidity is high is practical, but some water must be removed from the air if the system is to provide comfortable air—that is, the outdoor air mut be treated to reduce the humidity. To accomplish this, the compressor system is allowed to operate even though it is not needed for cooling. When the outdoor air moves across the much cooler evaporator coils, the air will be further chilled, causing water to condense from it. The cooled air is then fed inside the building, where it mixes with warmer air. The combined air is cool and low in humidity.

The circuit that provides system control when the outdoor air temperature is below 60°F and humidity is high requires only the addition of the humidity sensor contacts (numbered 8 in Figure 8–7). They are normally open contacts, which close when the humidity exceeds a preset limit.

When there is high humidity, relay *K*103 will close every time there is an output from the *Y*1 terminal of the thermostat (that is, a call for cooling). The circuit to the coil of relay *K*103 is through the contacts of the humidity sensor. The damper system will operate as before, with the damper motor rotating to open and close the damper. The mixed air temperature sensor continues to control the direction of motor rotation. The dampers modulate, keeping the air temperature at the mixed air temperature sensor around 60°F.

Outdoor Air Temperature Between 60°F and 70°F

If the outdoor air temperature is between 60°F and 70°F, the air is still useful for cooling purposes but is not sufficiently cool to satisfy all cooling needs—this is assuming that the return air will have a temperature above 70°F. Such a situation is normal when cooling is called for in most systems.

In order to make use of the available cooling from outdoor air, the outdoor damper should be fully open, the return damper fully closed, and the exhaust damper fully open. With the dampers in the indicated positions, the maximum amount of outdoor air (below 70°F) will be admitted to the system. The warmer return air will be blocked by the closed return damper. The warm return air will be discharged to the outdoors, through the exhaust damper. The circuit connection for accomplishing the required damper movements is shown in Figure 8–10.

Since the outdoor air temperature is between 60°F and 70°F, the outdoor air > 70°F thermostat will not activate. The contacts of the outdoor air temperature below 60°F thermostat will be in the position shown as 1 in Figure 8–10. The mixed air temperature is above 60°F. The mixed air temperature below 60°F thermostat will not activate. A complete circuit exists to terminal 3 of the damper motor. The motor will rotate to the position where the outdoor damper is open, the return damper closed, and the exhaust damper open. Any further action of the air-conditioning system should not cause a change in the dampers.

Relay *K*101 Figure 8–10 is de-energized. (This relay is numbered in Figure 8–7.) The 1 and 2 contacts of relay *K*103 (numbered 9 in Figure 8–7) will be closed. Relay *K*103 will energize every time there is a call for cooling from the thermostat. The closing of the contacts of relay *K*103 thereby provides for a complete circuit to the compressor contactor coil (see Figure

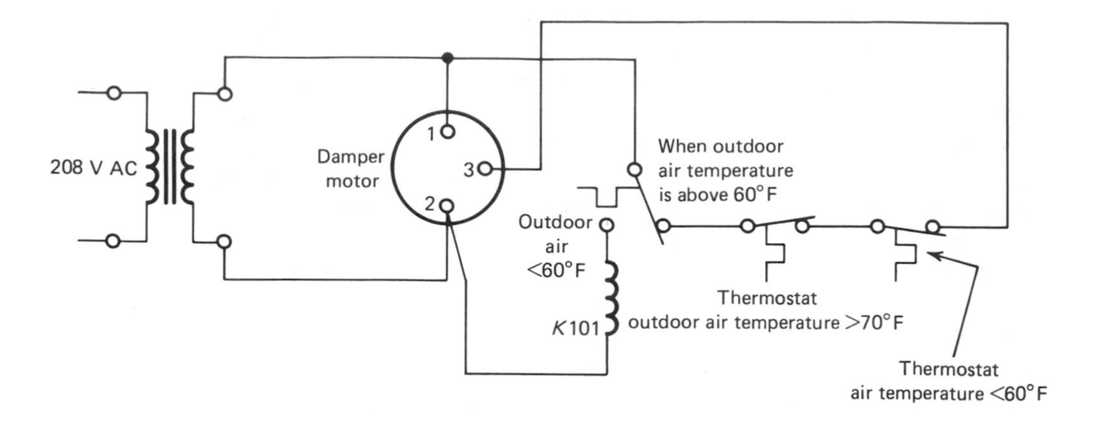

Figure 8–10. Outdoor air temperature between 60°F and 70°F.

6–16). The compressor will operate whenever there is a call for cooling from the thermostat.

Outdoor Air Temperature Above 70°F

Whenever the outdoor air temperature is above 70°F, it is considered (in this energy saver system) to be unsuitable for cooling purposes. The desired damper conditions are outdoor damper closed, return damper open, and exhaust damper closed. These conditions are met by the action of the contacts associated with the outdoor air temperature greater than 70°F thermostat. These contacts are shown as 2 in Figure 8–10. When the outdoor air temperature is above 70°F, the thermostat contacts will open. This action will interrupt the circuit to 3 of the damper motor. The motor will rotate, closing the outdoor damper. The return damper will open, and the exhaust damper will close.

The compressor will operate whenever there is a call for cooling. Relay $K101$ is de-energized, and its 1 and 2 contacts are closed. The circuit to the coil of relay $K103$ is complete. It will energize whenever the thermostat calls for cooling. The 1 and 3 contacts of $K103$ are closed, completing the circuit to the compressor contactor coil.

Electronic Control of Damper Systems

A number of electronically controlled damper systems are presently being developed. The use of electronic control allows more sensitive sensing of conditions.

With a thermostat that operates a set of switch contacts mechanically, the differential of turn on and turn off may be 3 or 4 degrees Fahrenheit. This differential may be reduced by using an electronic sensor and amplifier system.

There are control units presently available that sense return air temperature and humidity and compare it with outdoor air temperature and humidity. The selection of outdoor air or return air is based on which contains less *total* heat.

The introduction of new electronic control systems is proceeding at a rapid pace. Standardization of these control systems may take place in the future. For the present, each system must be studied individually.

Review Questions

Chapter 8

Mark each statement T for true or F for false.

1. In most air conditioning systems, the dampers operate independently of each other. 1. __
2. The exhaust damper provides a path for the mixed air. 2. __
3. The outdoor damper is closed tight whenever the outdoor air temperature is above 65°F. 3. __
4. If the outdoor air temperature is below 60°F, the dampers are expected to modulate. 4. __
5. When outdoor air alone is used for cooling, the outdoor damper should start to close when the mixed air temperature goes below 60°F. 5. __
6. The return damper never closes. 6. __
7. It is normal to have the outdoor damper open slightly whenever the system is in operation. 7. __

Workbook Section

Complete the individual diagrams as directed in the instructions given.

1. Connect the motor to input lines L1 and L2. The motor is to operate when power is available between L1 and L2.

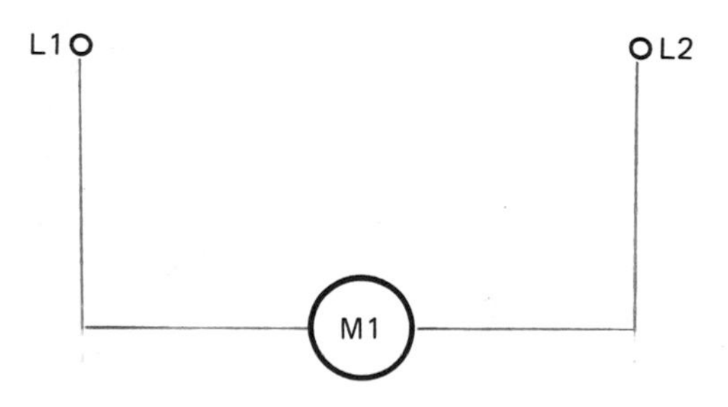

2. Connect the motor and switch in a circuit to L1 and L2. The motor is to operate when the switch is closed.

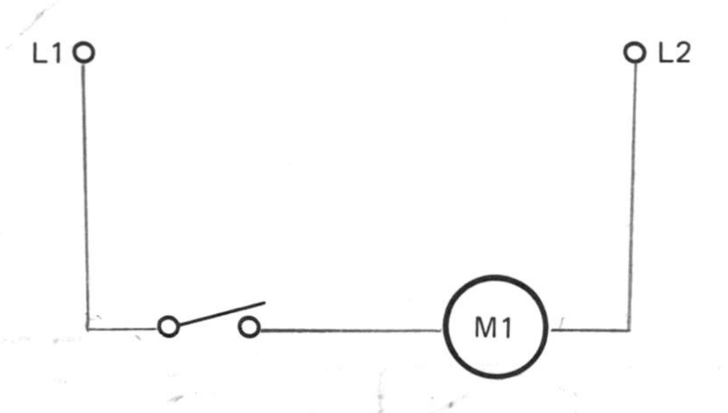

3. Connect the circuit so that when switch S1 is closed, motor M1 will operate, and when switch S2 is closed, motor M2 will operate.

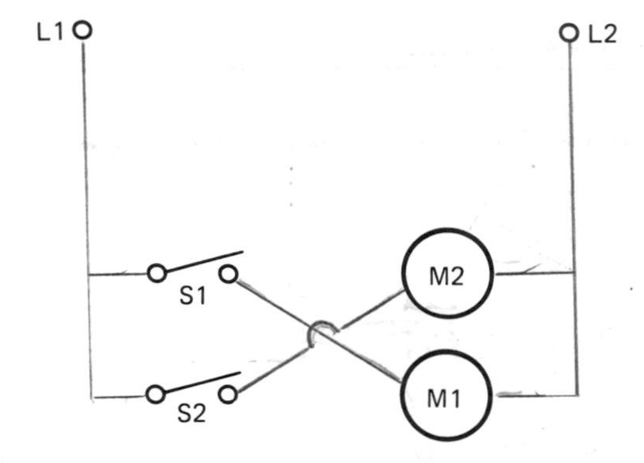

4. Connect the circuit so that when switch S1 is closed, motors M1 and M3 will operate. Motor M2 is to operate when switch S2 is closed.

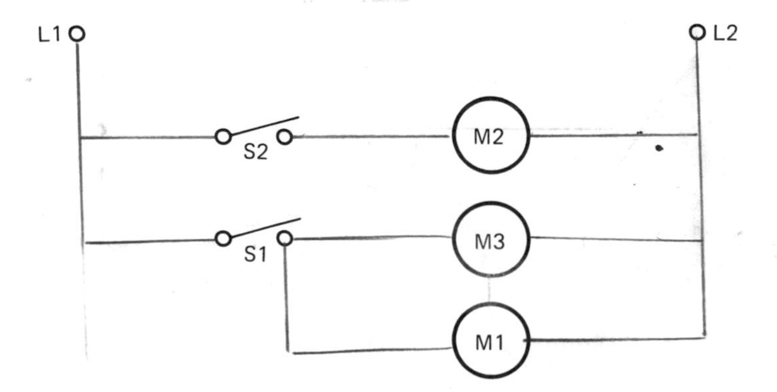

5. Connect the circuit so that when switch S1 is closed, motors M1 and M3 will operate. When pressure switch PS1 opens, motor M3 is to stop. Motor M2 is to operate when switch S2 is closed.

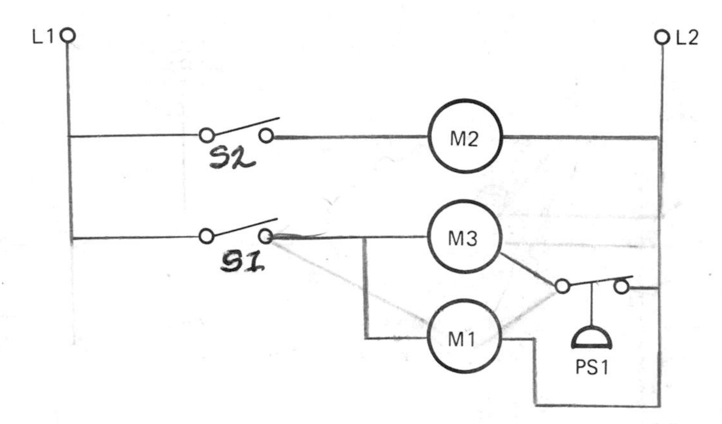

6. Connect the circuit so that the room thermostat controls the compressor contactor coil and the condenser fan contactor coil. The fan switch is to control the insider blower contactor coil.

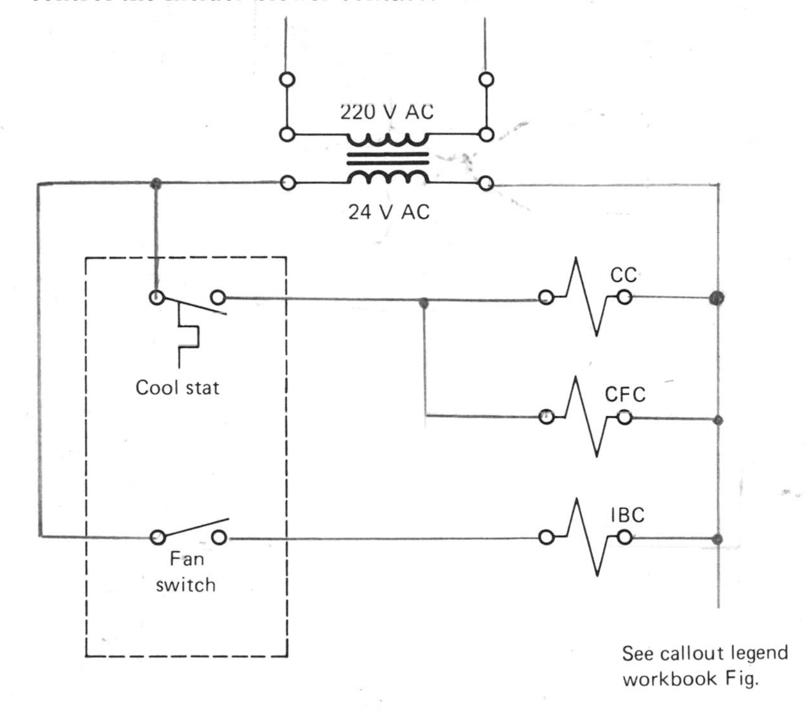

See callout legend workbook Fig.

7. Connect the controls in the proper circuit. The inside blower is to operate when either heat or cool is called for by the thermostat. No feedback is allowed through thermostat output *G* when in the "heat" position.

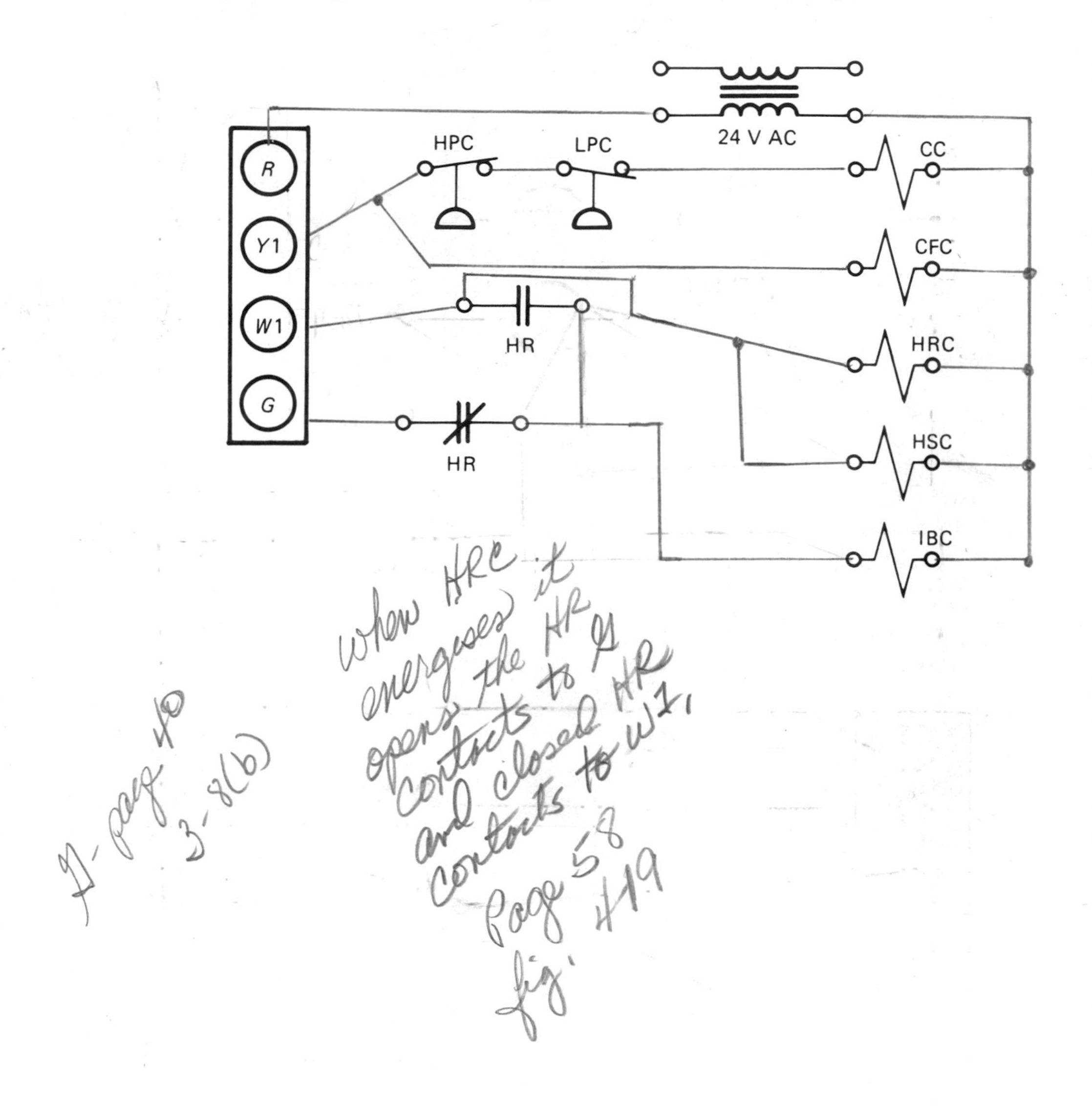

8. Connect the high-voltage components to L1 and L2. The 24-volt contactor coils are controlled by the appropriate thermostat output. The HPC and LPC connect in series with the compressor contactor (CC) coil.

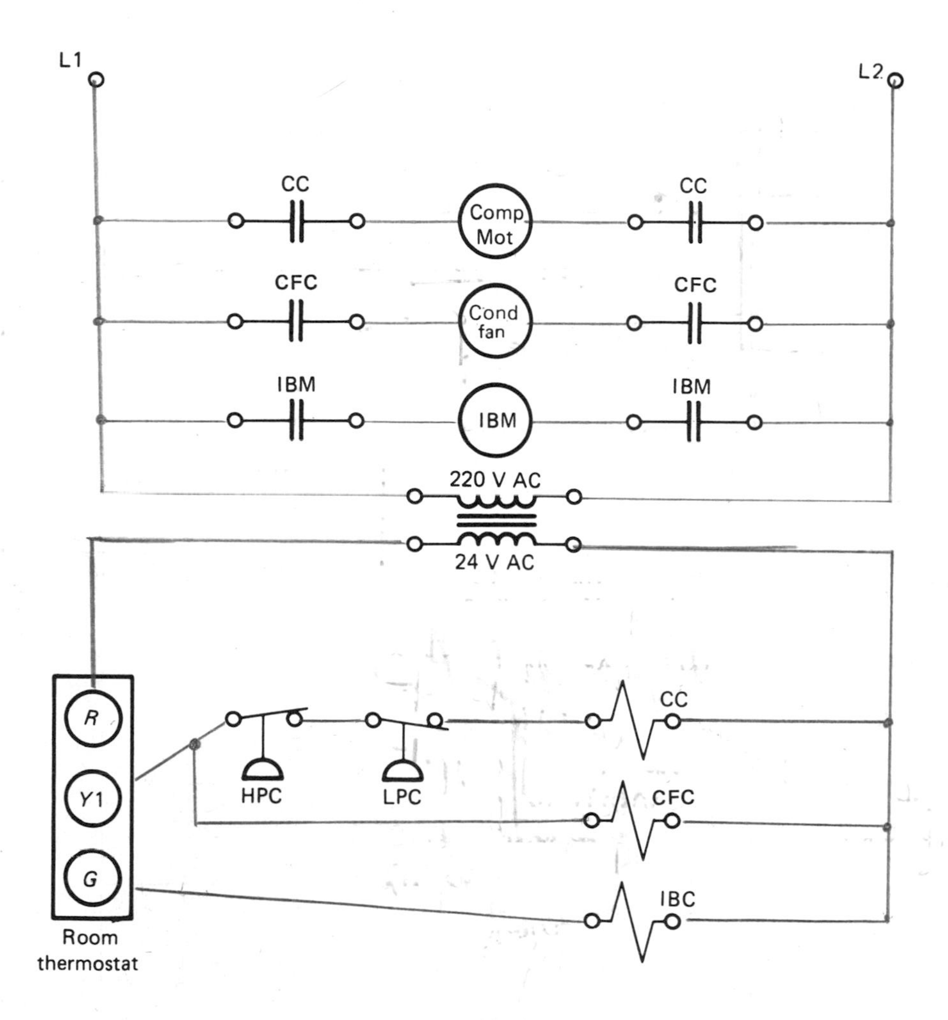

9. Complete the ladder diagram of the low-voltage circuit of a two-stage heat, two-stage cool air-conditioning system. The inside blower is to operate in both heat and cool. Compressor 2 is delayed by the time-delay relay.

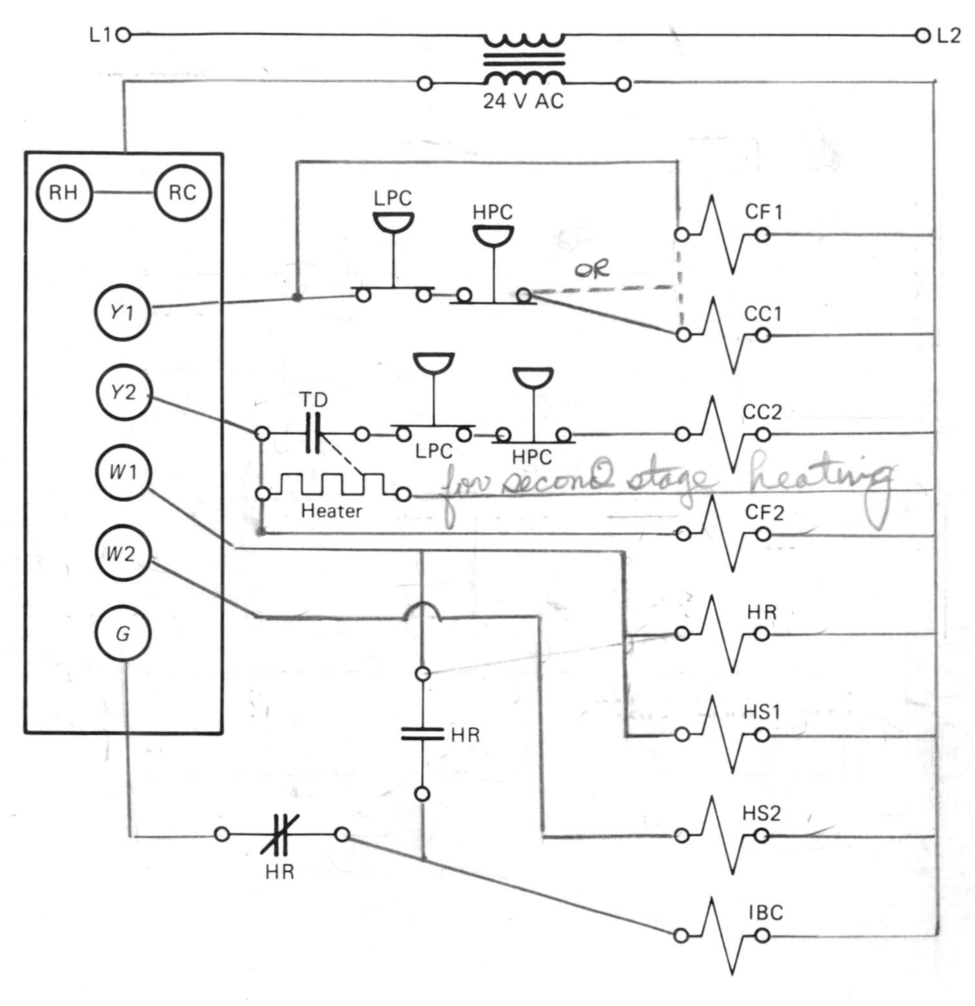

Legend

CC	Compressor contactor	HS1	Heat strip 1 contactor
IBM	Inside blower motor	HS2	Heat strip 2 contactor
HR	Heat relay	CF1	Condenser fan 1 contactor
HS	Heat strip contactor	CF2	Condenser fan 2 contactor
IBC	Inside blower contactor	HR	Heat relay
HPC	High-pressure control	IBC	Inside blower contactor
LPC	Low-pressure control	LPC	Low-pressure control
CC1	Compressor 1 contactor	HPC	High-pressure control
CC2	Compressor 2 contactor	TD	Time-delay relay

10. Connect the components of the simple refrigerator in a schematic diagram.

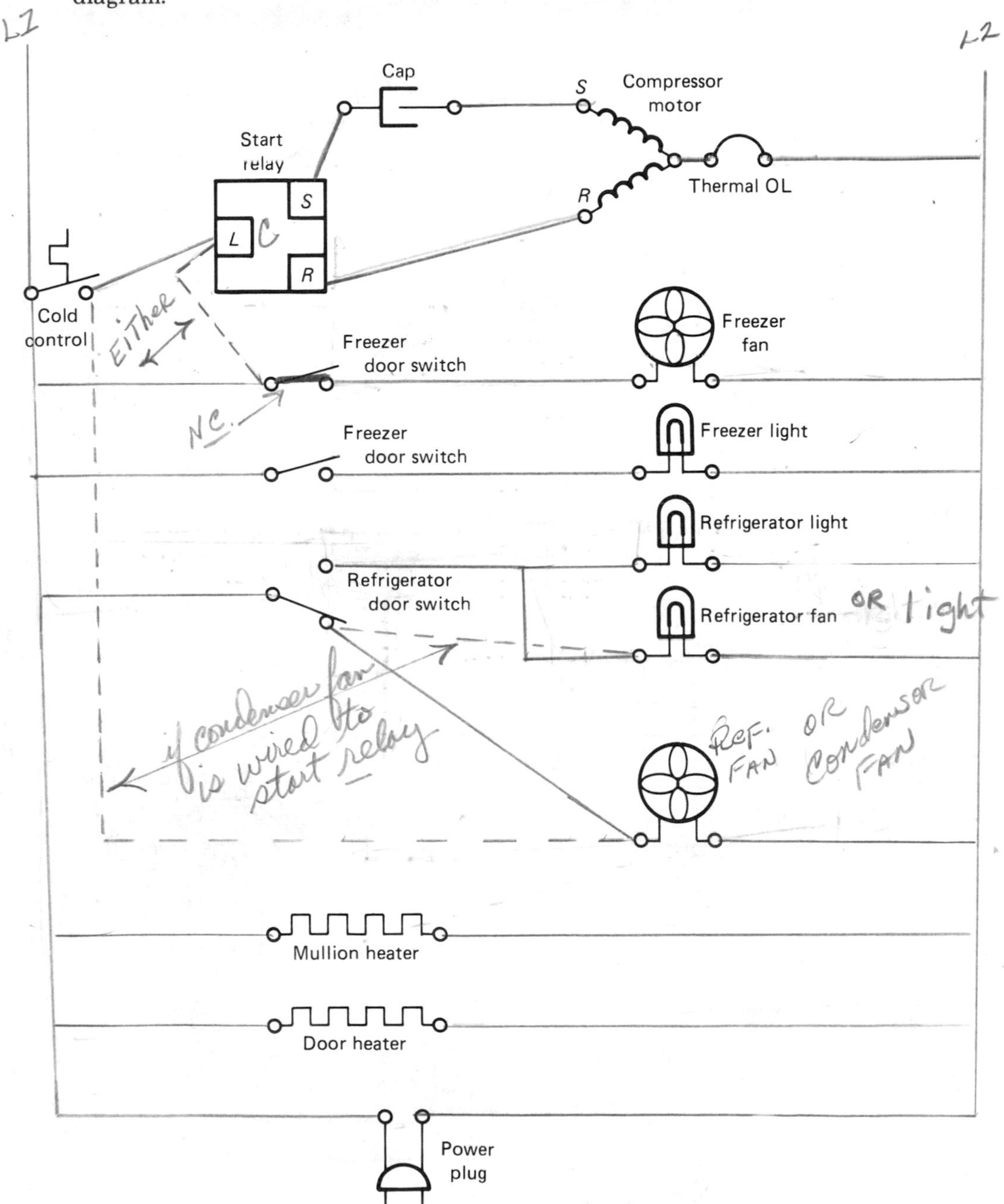

11. Connect the components of the frost-free refrigerator. The defrost timer circuit is of the accumulated time type.

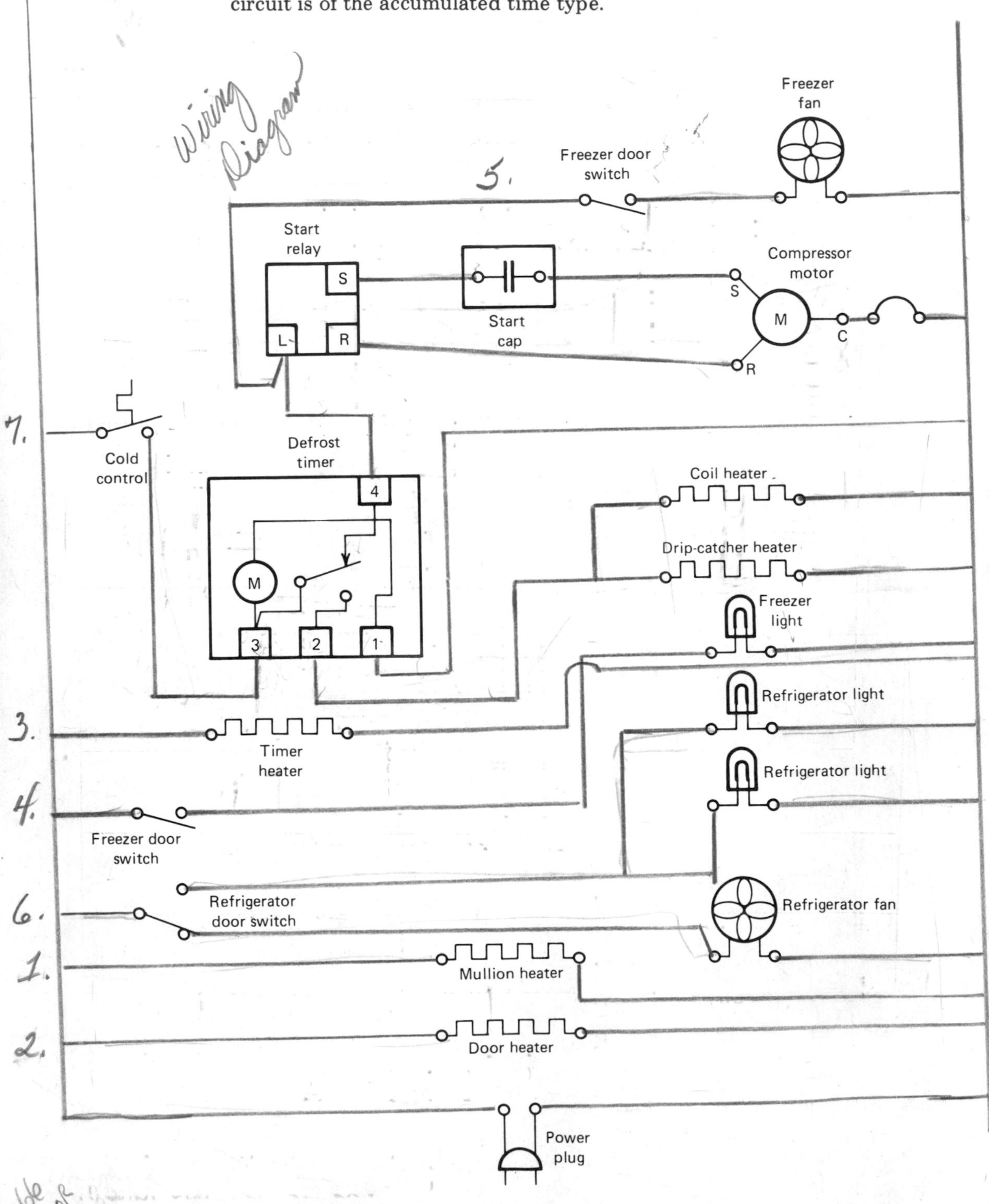

12. Connect the components to form a pictorial diagram of the frost-free refrigerator. Terminal strip tie posts may be added where necessary.

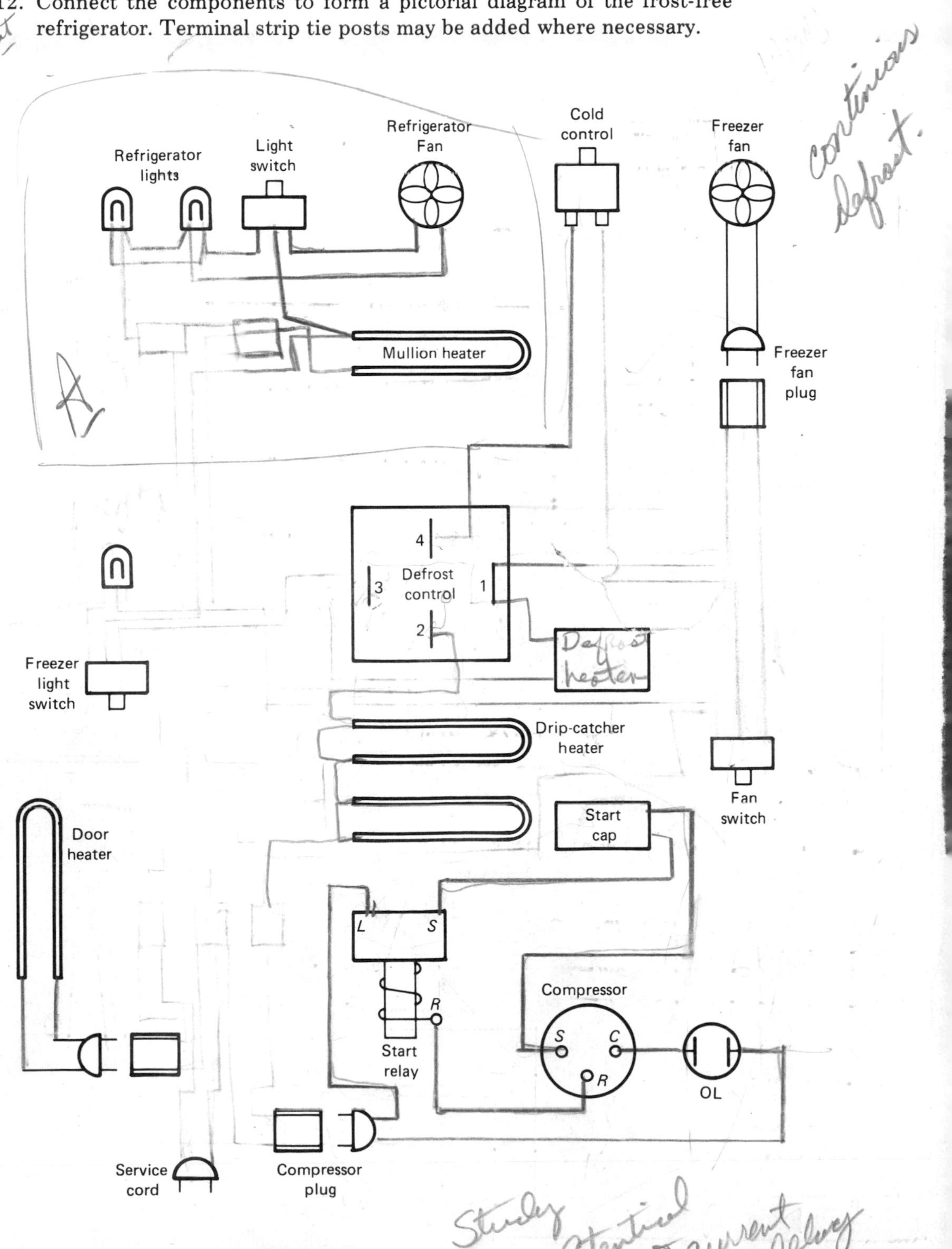

13. Connect the air-conditioning components to form a wiring diagram. The dual-pressure control is in the low-voltage control circuit. The overload is in the high-voltage control circuit of the compressor.

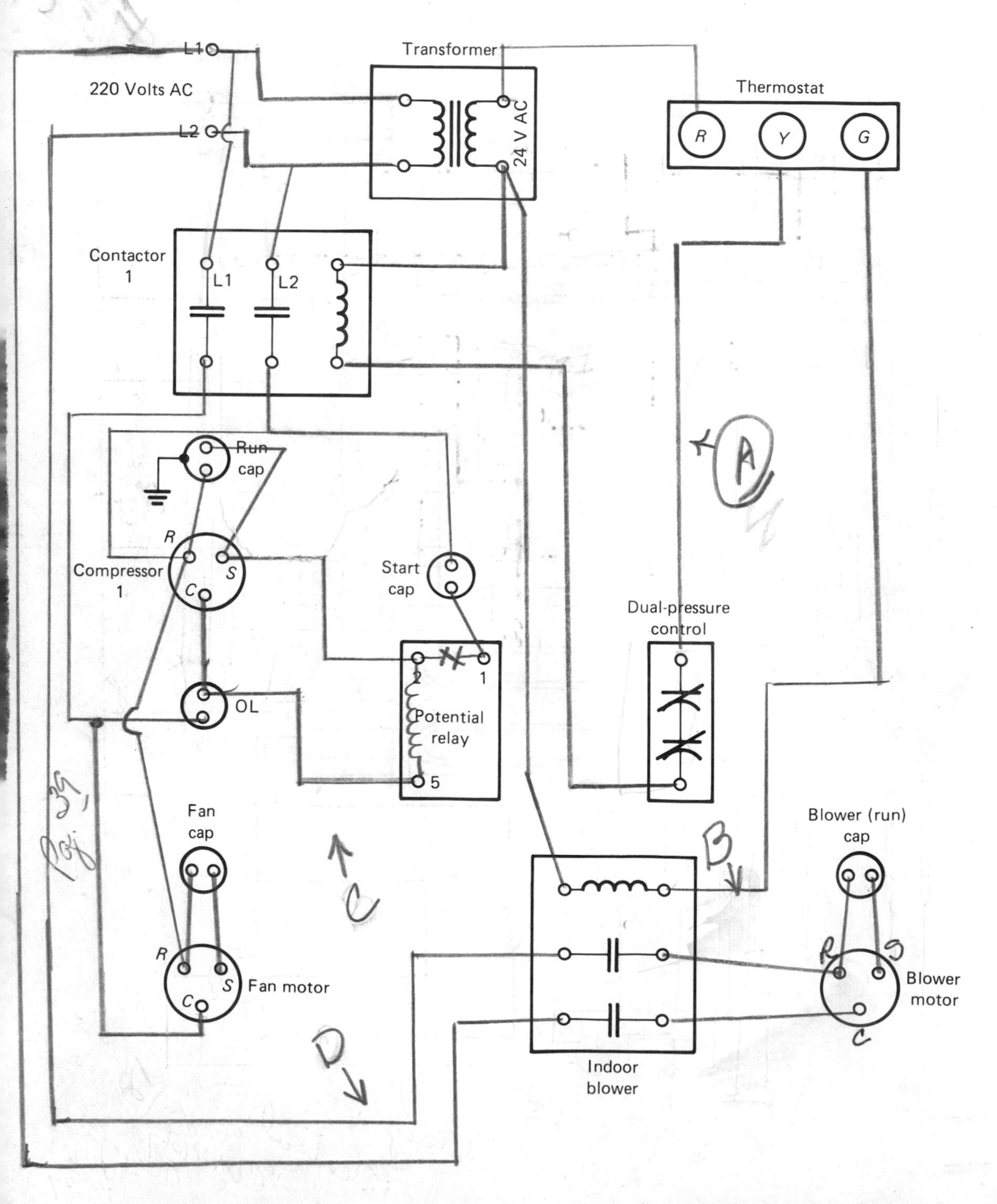

14. Connect the components to form a pictorial diagram. Power, 115 volts AC, is available between terminal strips TB1 and TB2. The refrigerator door switch controls the application of power to the lights and fan. The mullion heater is always connected across the input power.

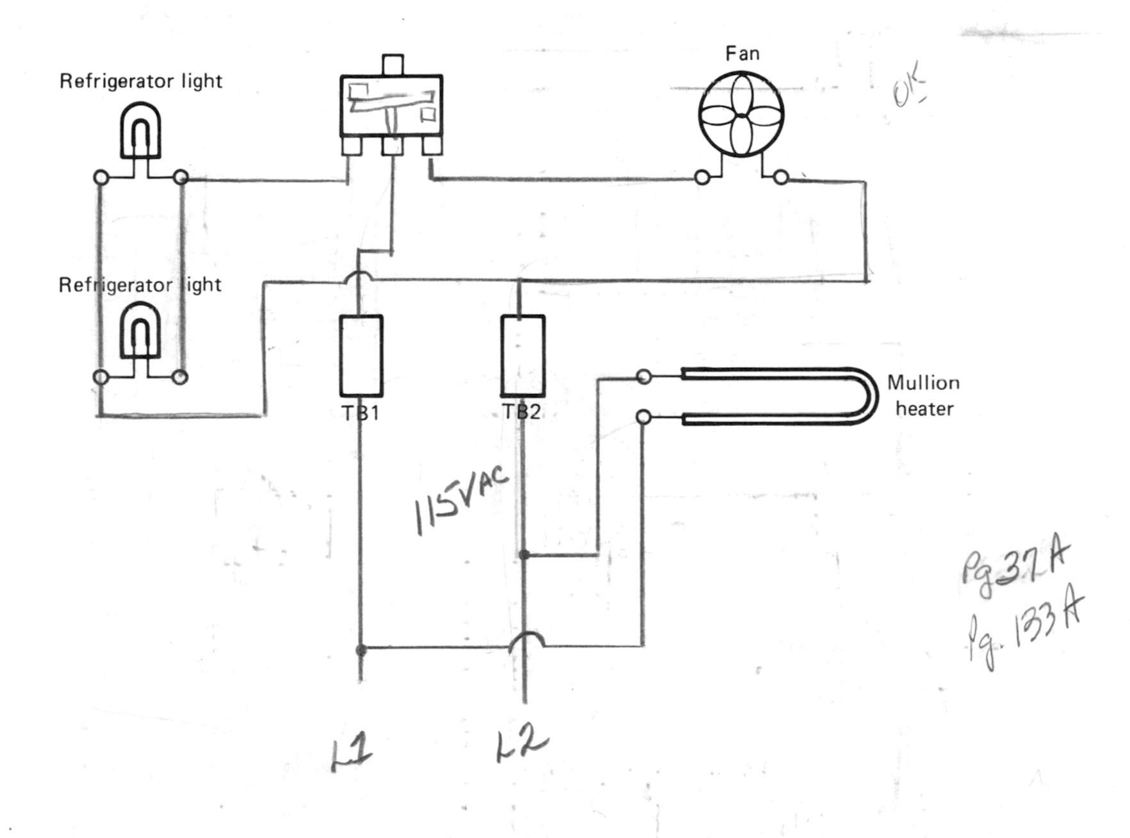

15. Connect the components of the air-conditioning evaporator unit to form a wiring diagram.

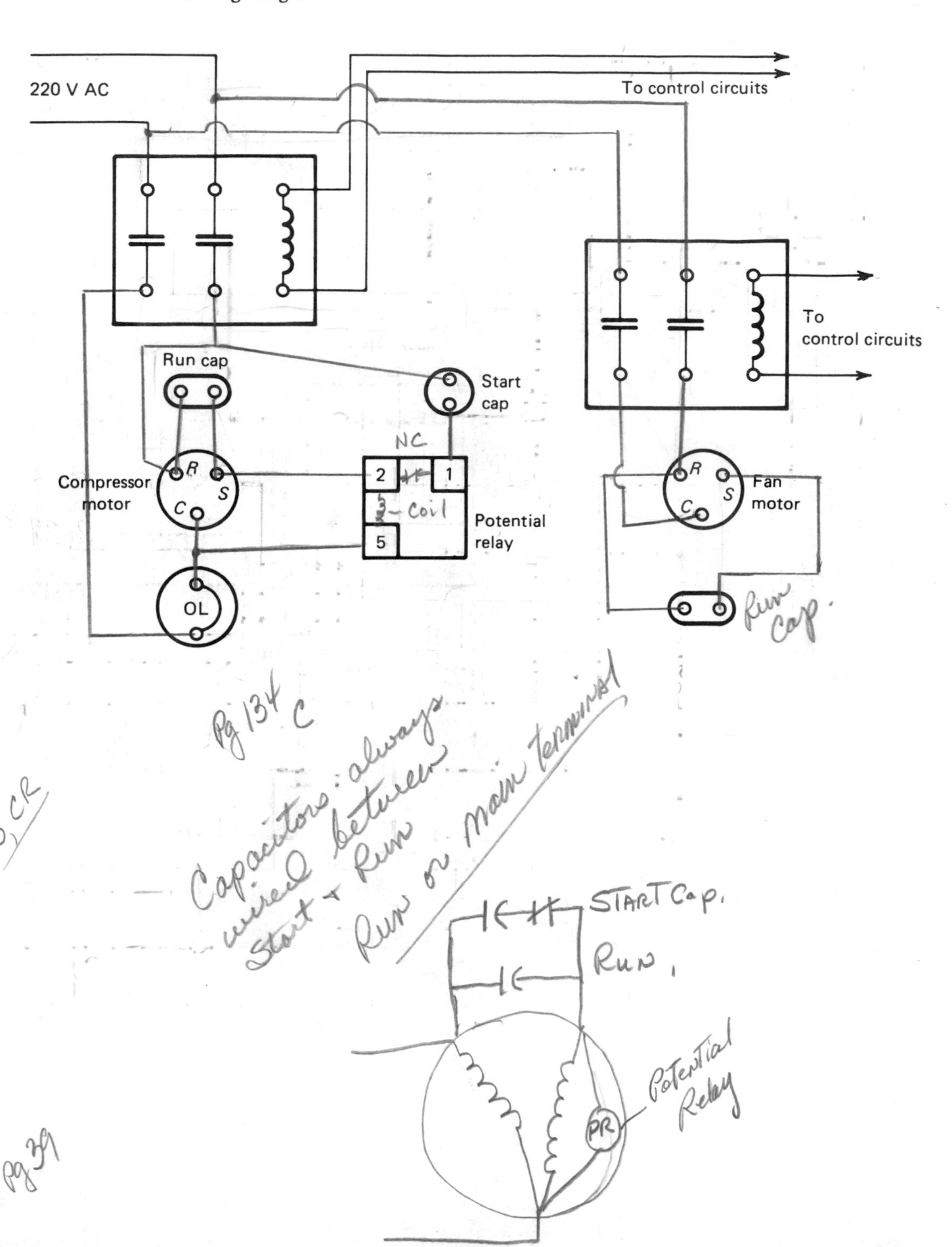

Quiz

16. Connect the components of the dual heat strip system in a wiring diagram. Terminal strips may be added where needed.

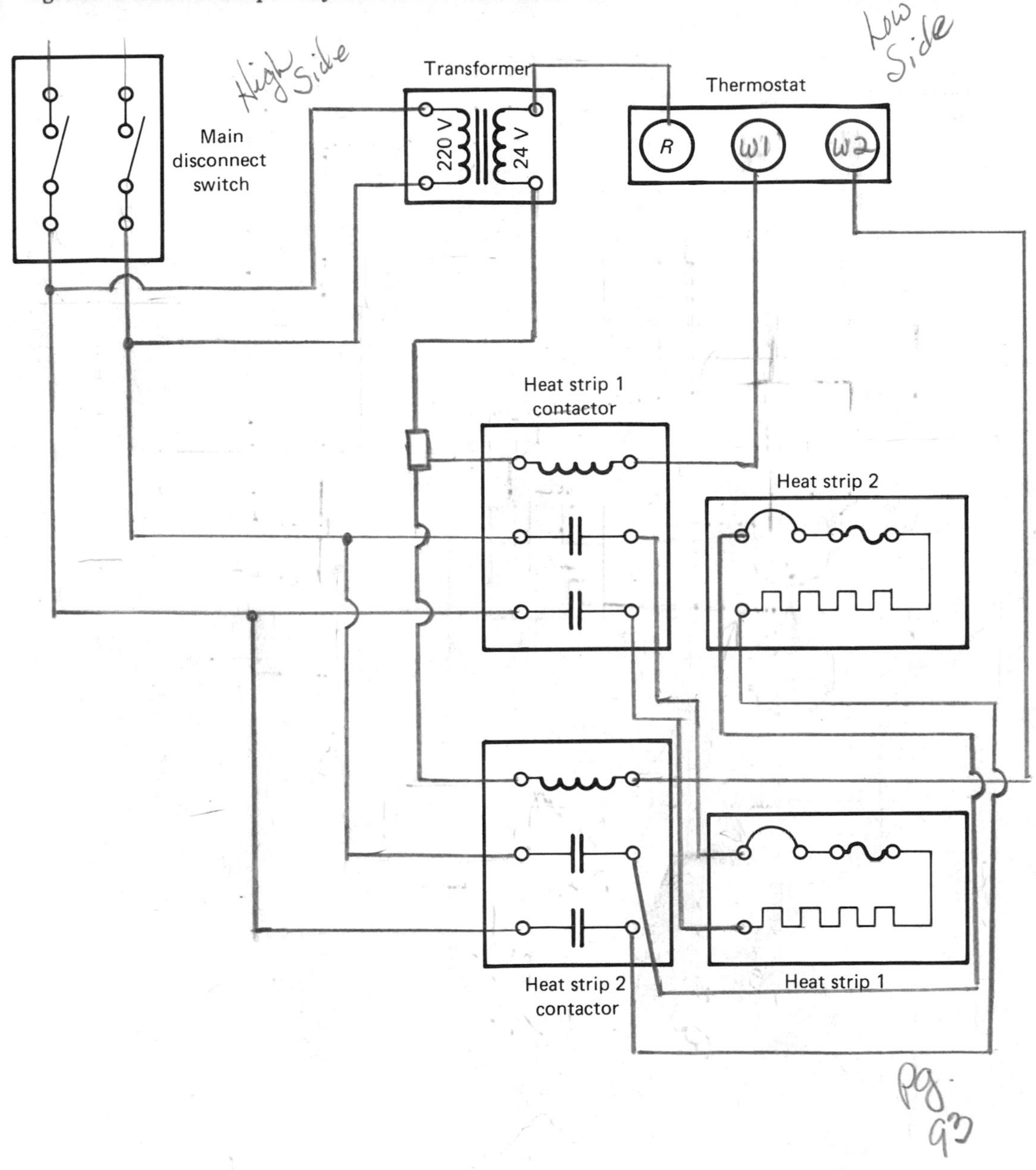

Pg. 93

17. Connect the components of the dual heat strip system to form a wiring diagram. Terminal strips may be added where needed. The heat relay is to be used to isolate the inside blower motor so that there is no feedback through the thermostat.

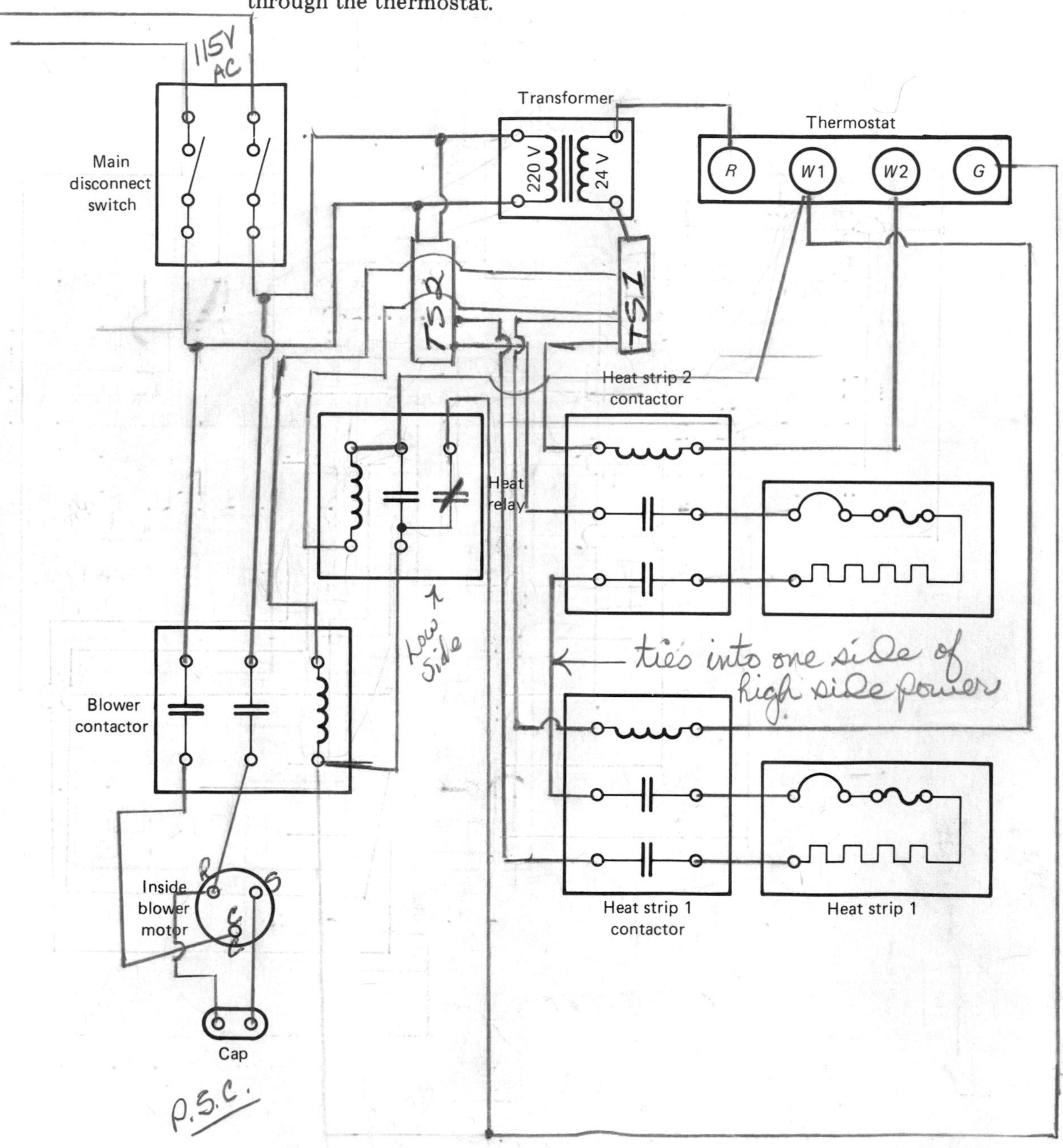

G-terminal has to go to N.C. contacts.

18. Connect the components of the dual heat strip system to form a wiring diagram. The heat relay isolates the inside blower control circuits. The heat strips and inside blower motor should be connected to best balance the loading effect.

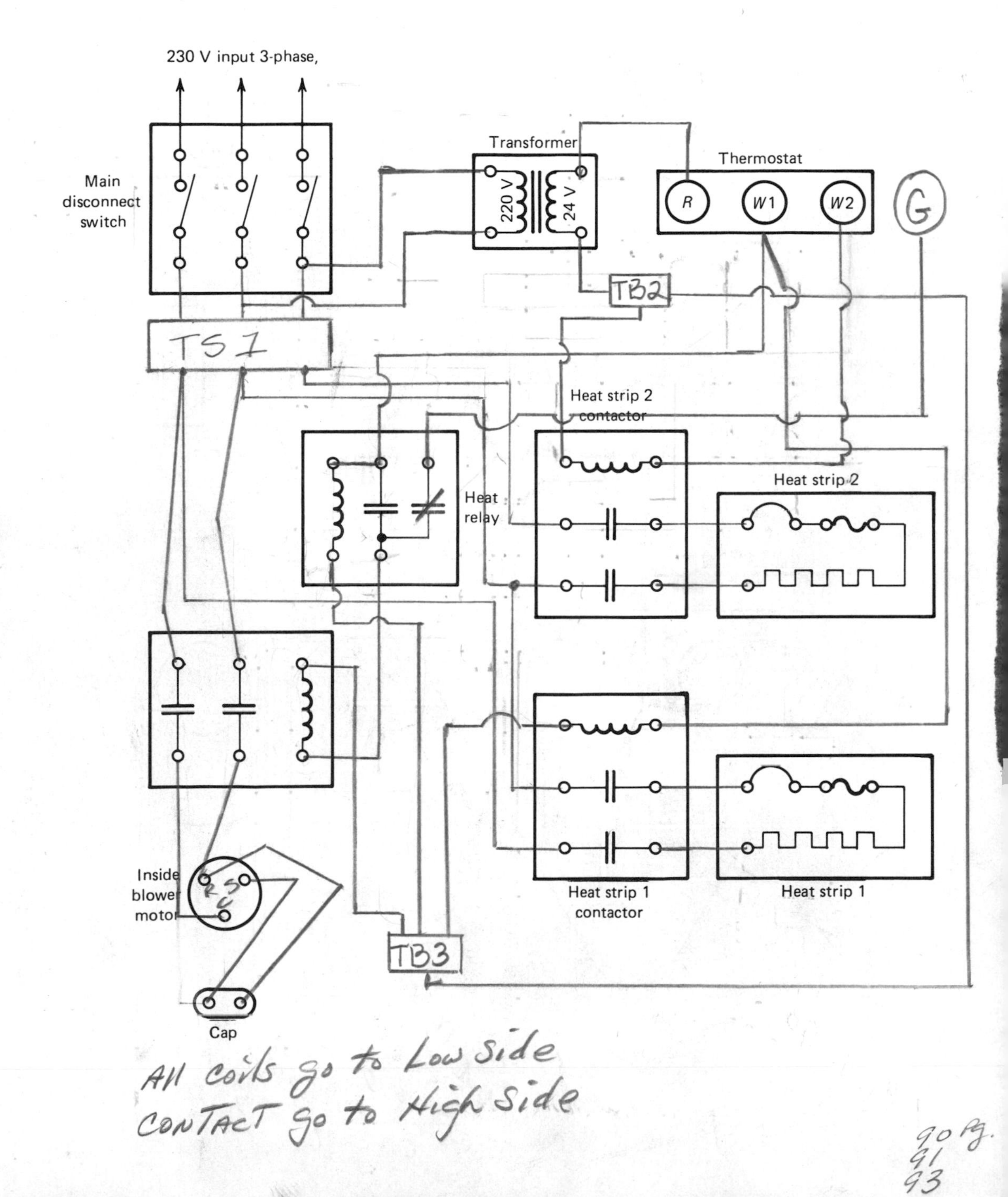

19. Connect the wiring diagram of the three-phase evaporator circuit including the low-voltage control of the evaporator.

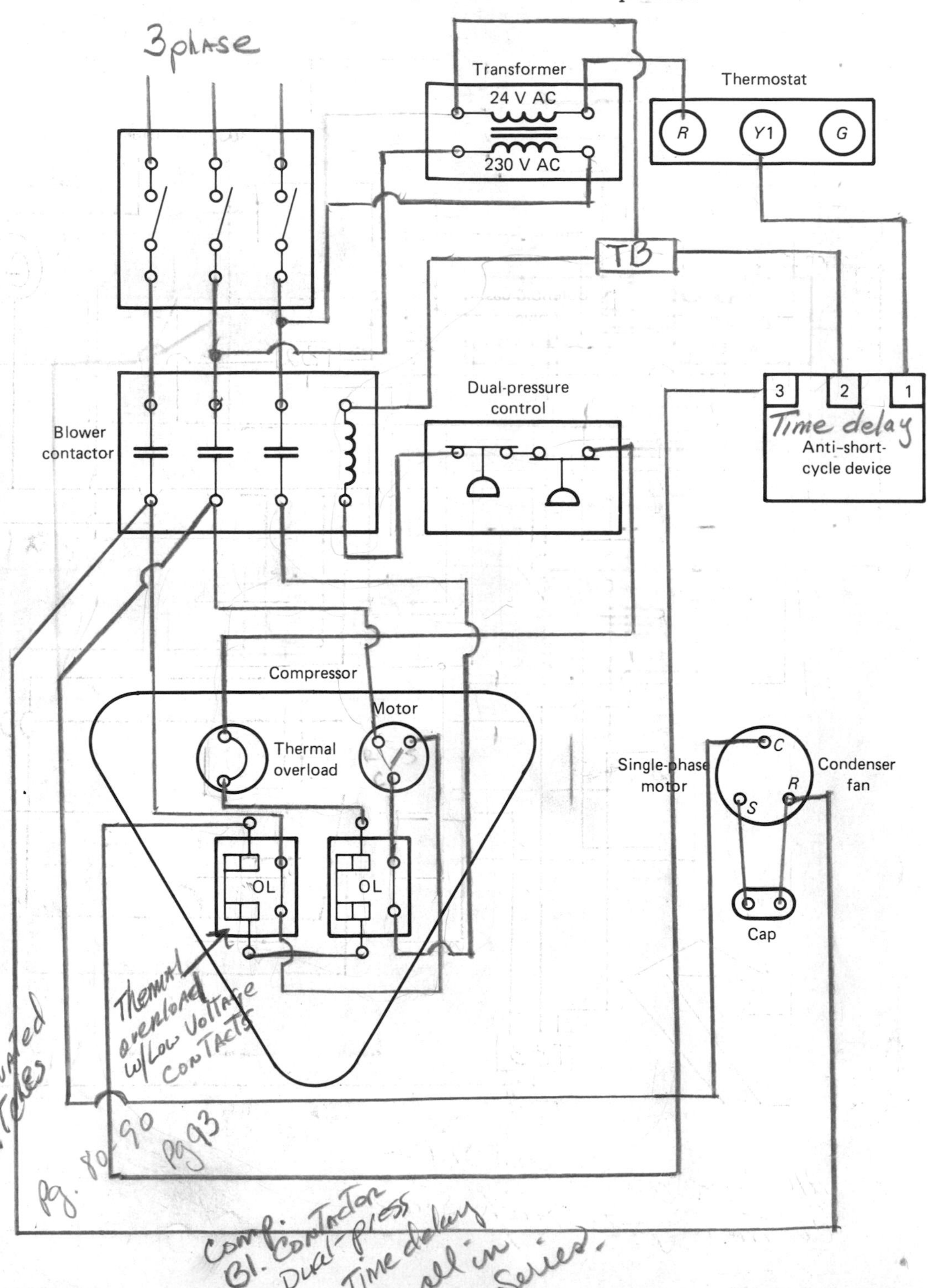

20. Connect the gas heating system. The fan relay is to be controlled from the G output of the thermostat.

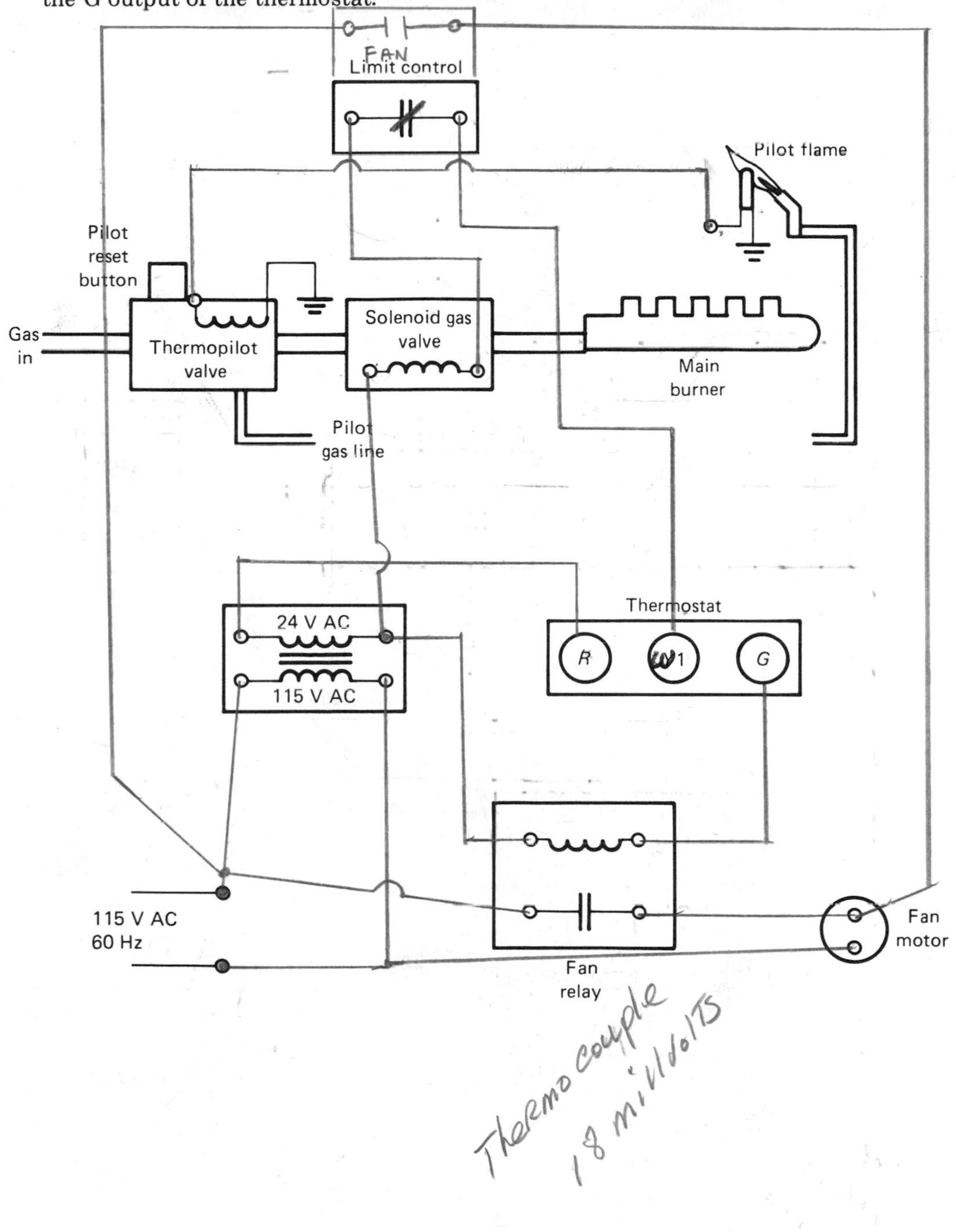

21. Connect the components of the energy-saving gas heating system.

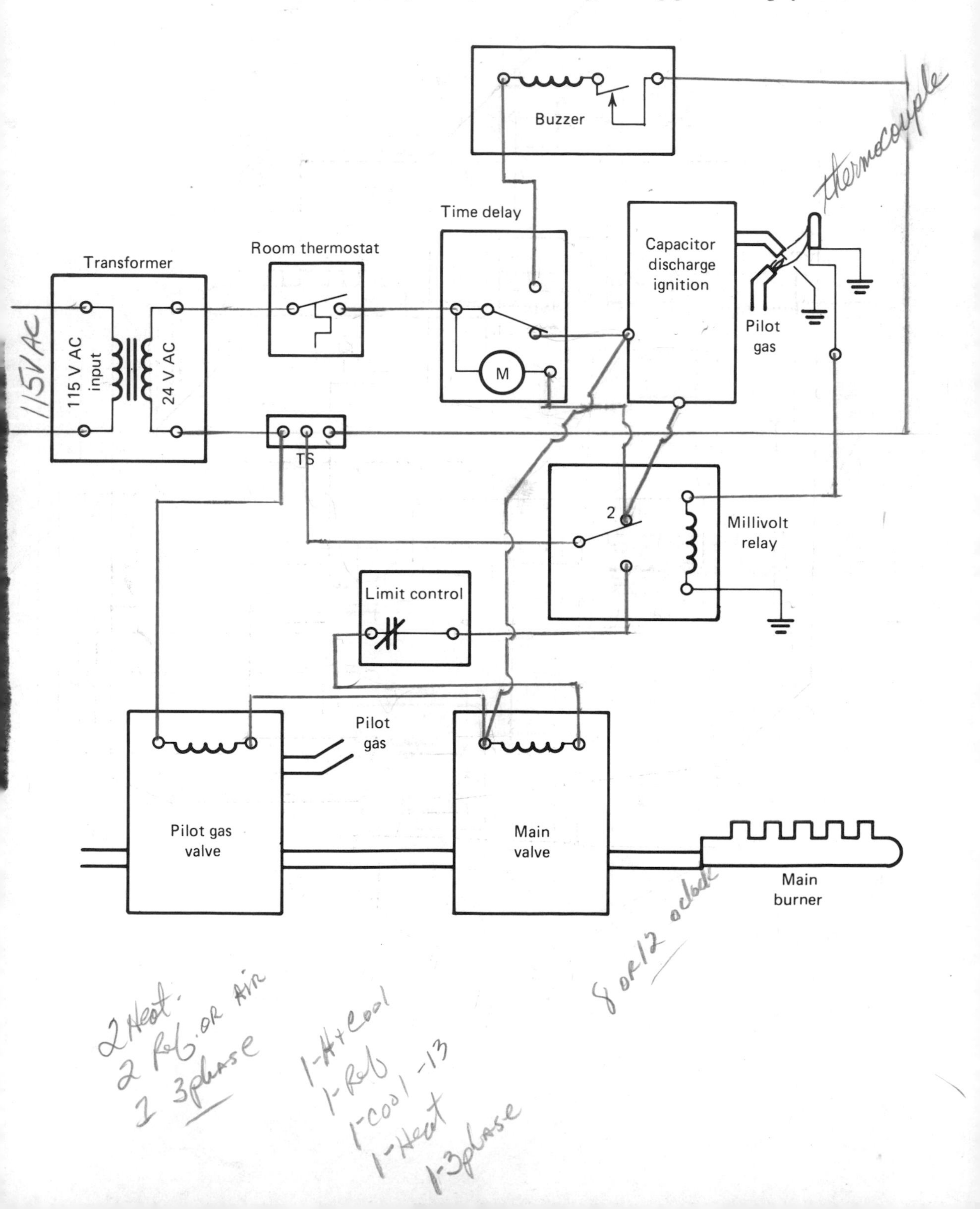

22. Connect the damper motor to the transformer output so that the damper will rotate to the closed position (outdoor air).

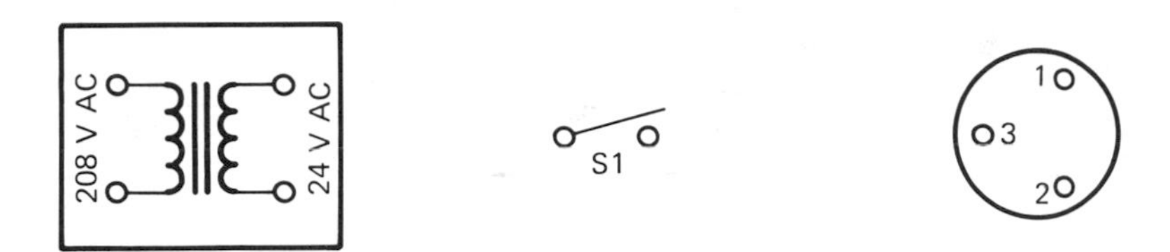

23. Connect the damper motor to the transformer output so that the damper will rotate to the closed position (outdoor air). When switch S1 is closed, the damper is to rotate to the open position.

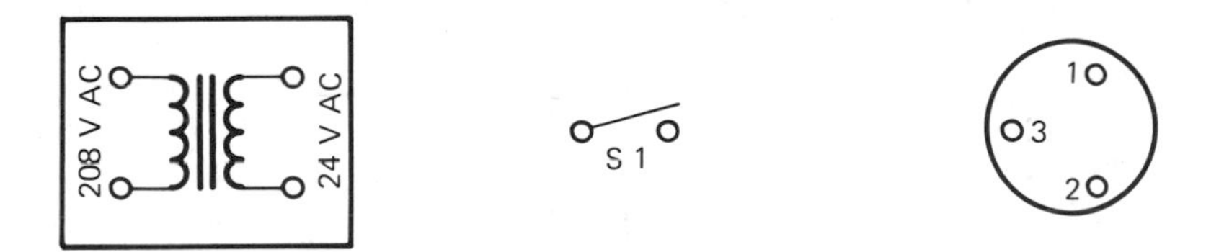

24. Connect the damper control circuit to provide for proper use of outdoor air for cooling purposes.

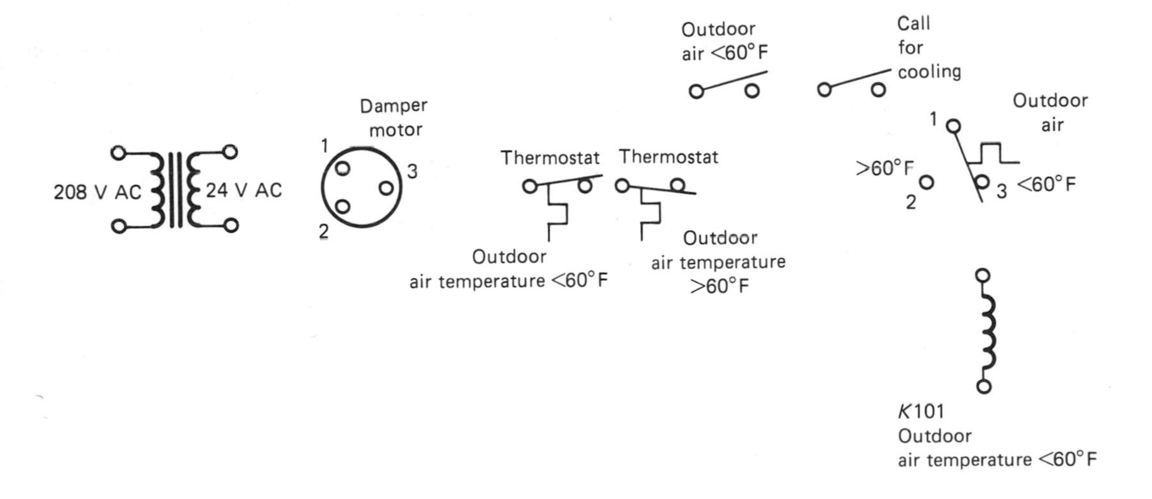

25. Connect the complete damper control system, including the energy-saving compressor shut-down system to be used when outdoor air warrants it.

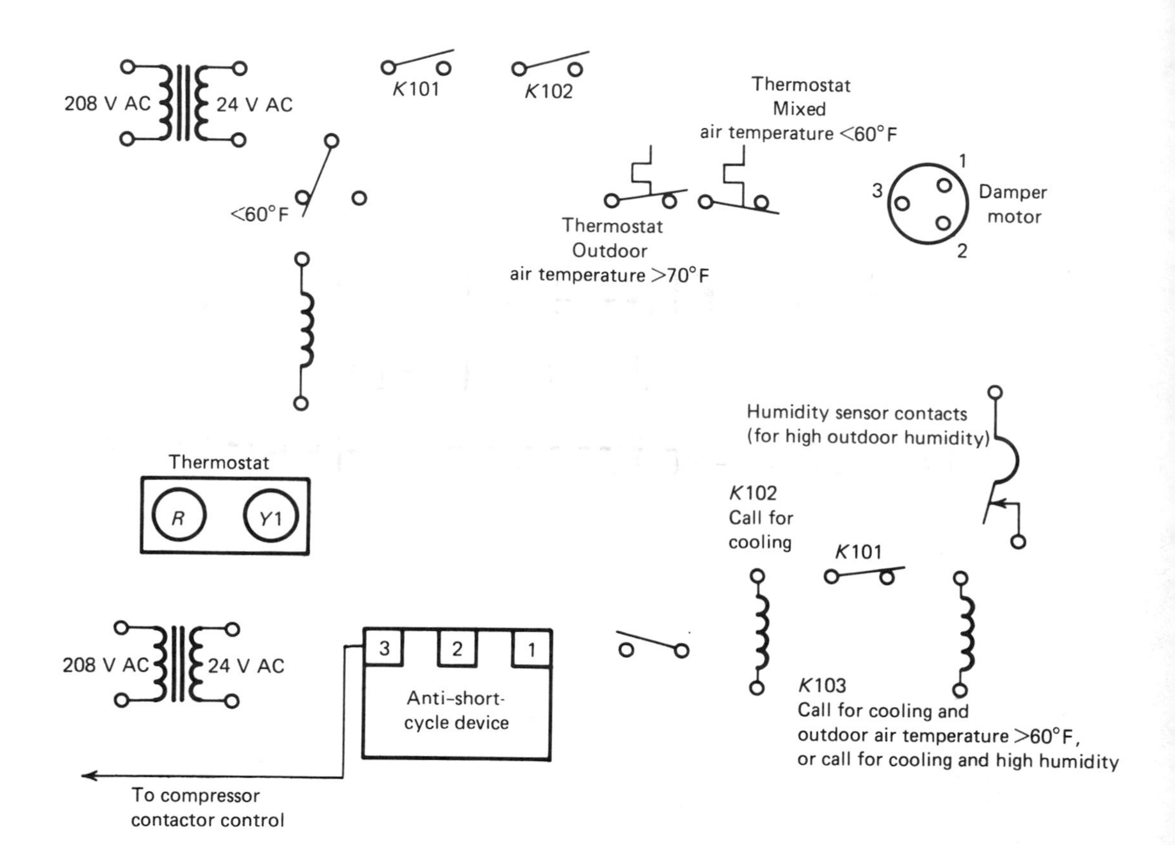

Appendix A

ELECTRICAL SYMBOLS COMMON TO AIR-CONDITIONING SYSTEMS

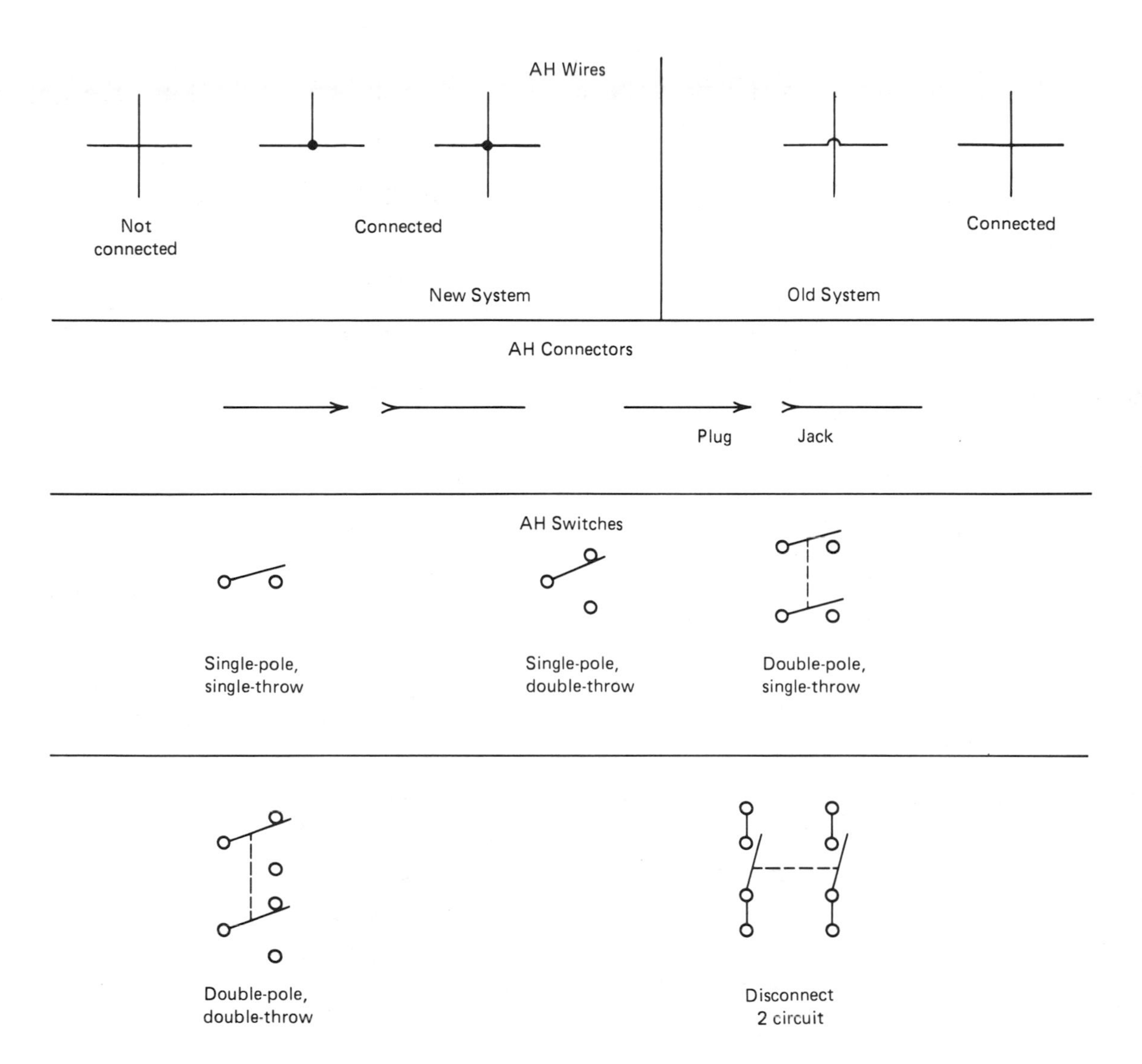
AH Wires
Not connected
Connected
New System
Connected
Old System
AH Connectors
Plug
Jack
AH Switches
Single-pole, single-throw
Single-pole, double-throw
Double-pole, single-throw
Double-pole, double-throw
Disconnect 2 circuit

Push-button Switch

Normally open

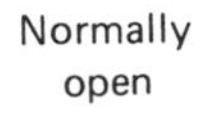

Normally closed

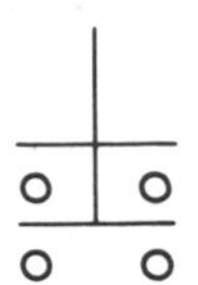

Double circuit, normally open

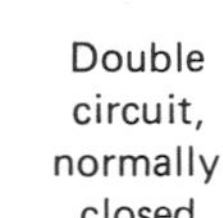

Double circuit, normally closed

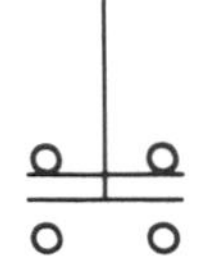

Double circuit, normally closed and normally open

Multi-position Switch

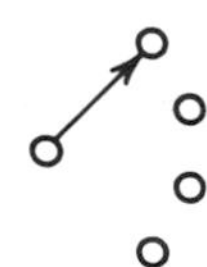

Rotary switch

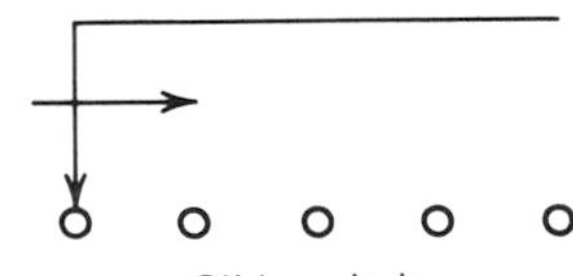

Slide switch

Limit Switch

Normally open

Normally closed

Foot Switch

Normally closed

Normally open

Liquid Switch

Normally open closes on level rise

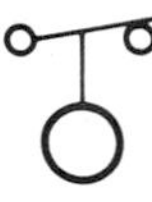

Normally closed opens on level rise

Flow Switch

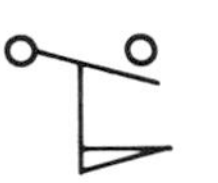

Normally open closes with flow

Normally closed opens with flow

Pressure Switch

Normally open, closes on pressure fall

Normally open, closes on pressure rise

Normally closed, opens on pressure rise

Normally closed, opens on pressure fall

Temperature Switch (Thermostat)

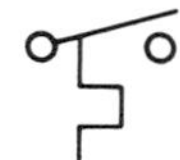

Normally open, closes on temperature fall

Normally open, closes on temperature rise

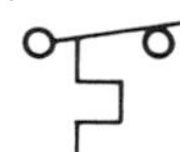

Normally closed, opens on temperature rise

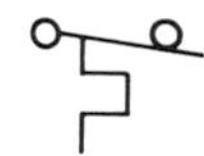

Normally closed, opens on temperature fall

Time-Delay Switch

Time-delay contact normally open, closes after delay

Normally open, closes after delay

Normally closed, opens after delay

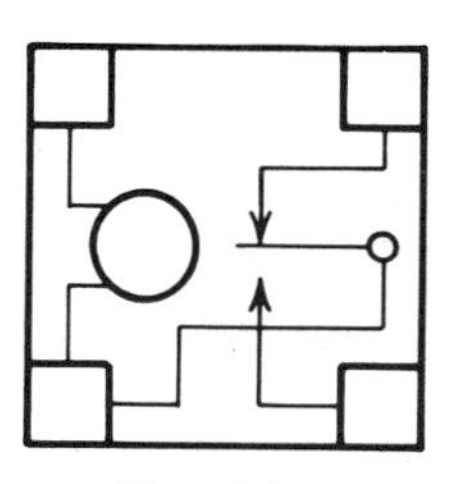

Time-delay motor

AH Contacts

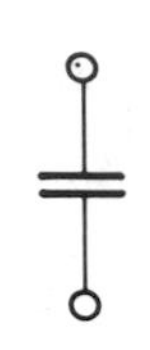

Normally open

Normally closed

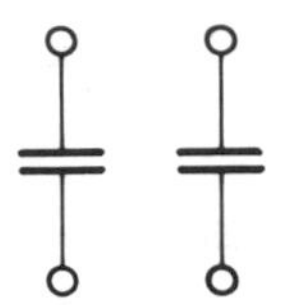

Double-pole, normally open

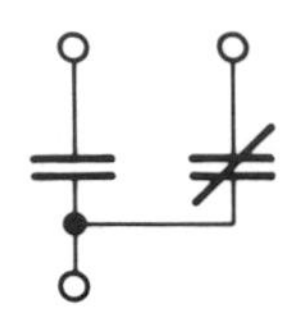

Double-pole, normally open and normally closed

AH Overloads

Thermal overload

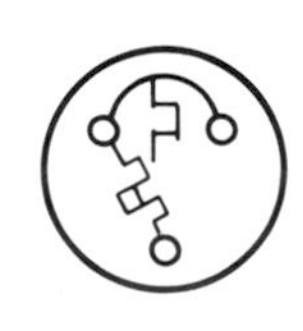

Thermal overload with heater

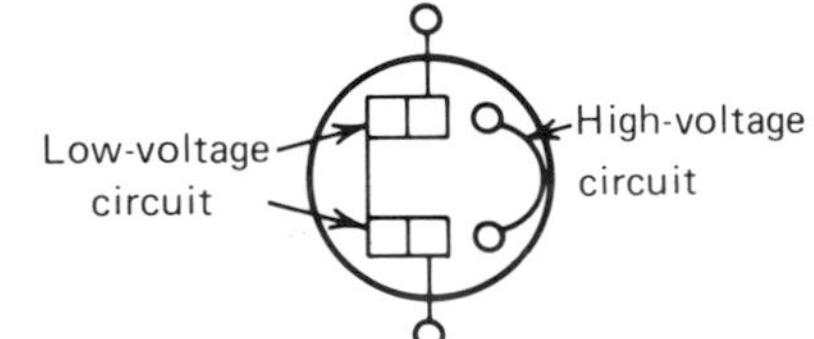

Thermal overload with low-voltage contacts

Fuse

Circuit Connections

Ground connection

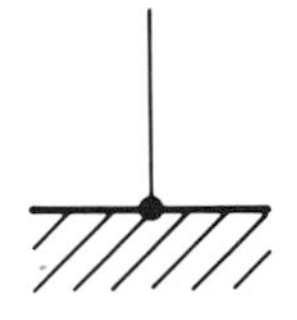

Chassis ground

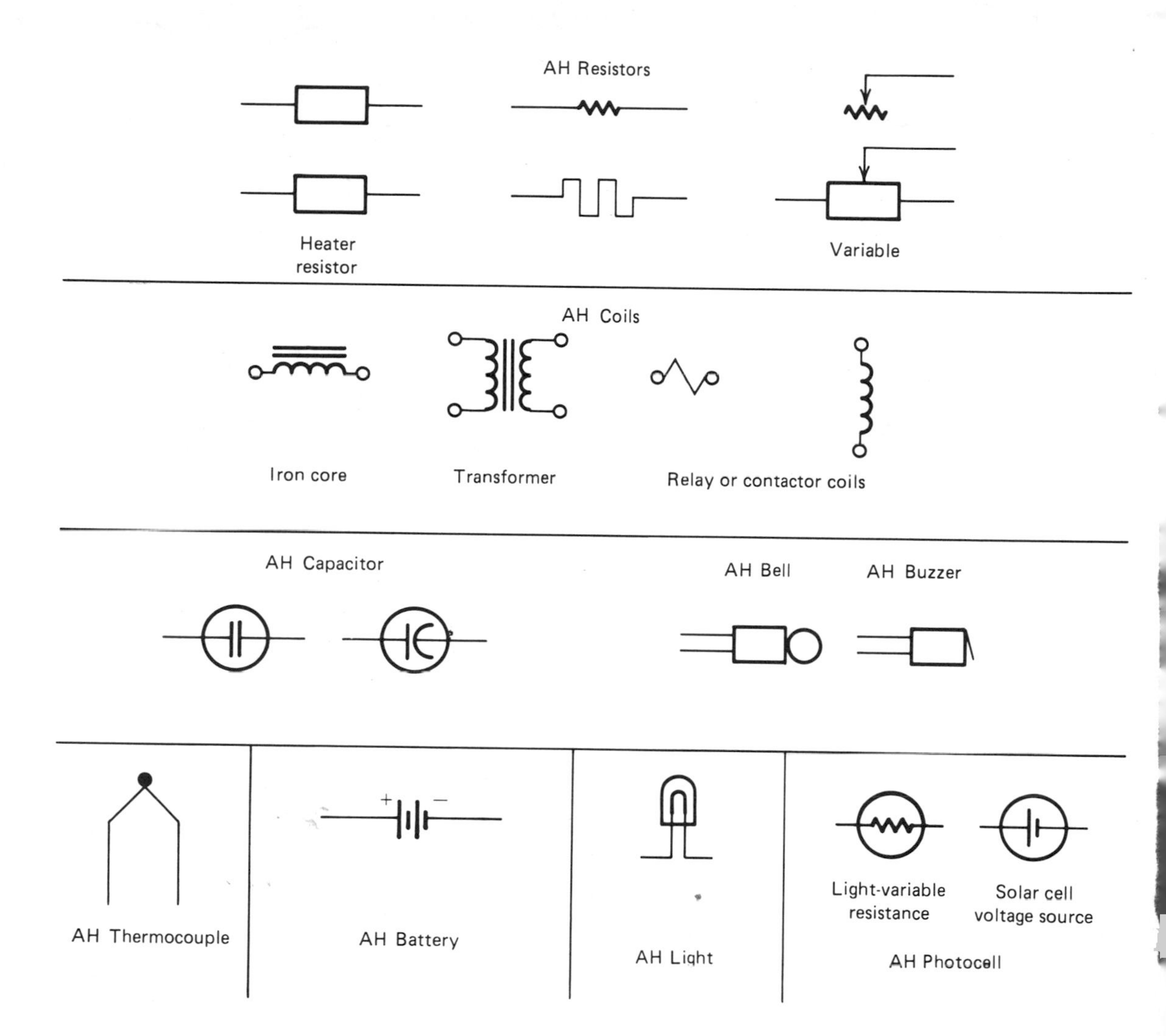
AH Resistors
Heater resistor
Variable
AH Coils
Iron core
Transformer
Relay or contactor coils
AH Capacitor
AH Bell
AH Buzzer
AH Thermocouple
+
−
AH Battery
AH Light
Light-variable resistance
Solar cell voltage source
AH Photocell

AH Motors

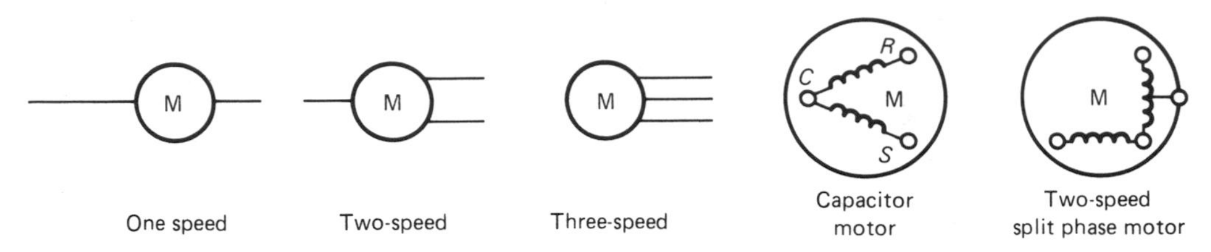

Three Phase Motor Connections

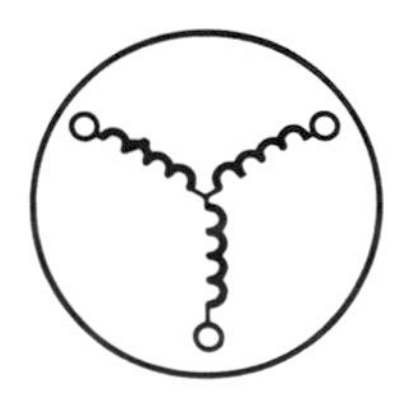

Three-phase wye

Three-phase delta

Appendix B

STANDARD VOLTAGES AND FREQUENCIES IN SELECTED COUNTRIES

MAIN VOLTAGES AND FREQUENCIES IN SELECTED COUNTRIES

Country	Voltage	Hertz
Afghanistan	220/380	50/60
Algeria	127/220	50
Argentina	220	50
Australia	240	50
Barbados	117	50
Bermuda	230/115	60
Bolivia	220/110	50/60
Brazil	220/127/110	50/60
Canada	230/115	60
Chile	220	50
Colombia	220/110	60
Costa Rica	220/110	60
Cuba	230/115	60
El Salvador	220/110	60
Equador	110/220	60
Haiti	230/115	60
Honduras	220/100	60
Hong Kong	200	50
Indonesia	220/127	50
Israel	230/400	50
Italy	220	50
Jamaica	220/110	50/60
Japan	100	50/60
Korea	210/105	60
Lebanon	190/110	50
Mexico	208/120	50
New Zealand	230	50

MAIN VOLTAGES AND FREQUENCIES IN SELECTED COUNTRIES *(continued)*

Country	Voltage	Hertz
Poland	380/220	50
Sweden	380/220	50
Switzerland	380/220	50
United Kingdom	240/230	50
U.S.A.	115/230–208/440	60
U.S.S.R.	220/127/110	50

Appendix C

MOTOR FULL-LOAD CURRENT

FULL-LOAD CURRENTS FOR THREE-PHASE 230-VOLT AC MOTORS

Motor Horsepower	Full Load (amps)	125 percent of Full Load (amps)
½	2.0	2.5
¾	2.8	~~2.5~~ 3.5
1	3.6	~~0.5~~ 4.5
1½	5.2	6.5
2	6.8	8.5
3	9.6	12.0
5	15.2	19.0
7½	22.0	28.0
10	28.0	35.0
15	42.0	52.0
20	54.0	68.0
25	68.0	85.0
30	80.0	100.0
40	104.0	130.0
50	130.0	162.0
60	154.0	192.0
75	192.0	240.0
100	248.0	310.0
125	312.0	390.0

Note: To obtain full-load currents for 208-volt motors, increase corresponding 230-volt, full load current by 10 percent.

FULL-LOAD CURRENTS FOR SINGLE-PHASE 230-VOLT AC MOTORS

Motor Motor horsepower	115 volts		230 volts	
	Full load (amps)	125 percent of Full Load (amps)	Full Load (amps)	125 percent of Full Load (amps)
1/6	4.0	5.5	2.2	2.8
1/4	5.8	7.2	2.9	3.6
1/3	7.2	9.0	3.6	4.5
1/2	9.8	12.2	4.9	6.1
3/4	13.8	17.2	6.9	8.6
1	16.0	20.0	8.0	10.0
1 1/2	20.0	25.0	10.0	12.5
2	24.0	30.0	12.0	15.0
3	34.0	42.0	17.0	21.0
5	56.0	70.0	28.0	35.0
7 1/2			40.0	50.0
10			50.0	62.0

Note: To obtain full-load currents for 208-volt motors, increase corresponding 230-volt motor full-load current by 10 percent.

Appendix D

WIRE SIZE REQUIREMENTS FOR LINE FEED TO MOTORS

WIRE TRADE NAME AND LETTER TYPE

Trade name of covering	Letter Code
Rubber	R
Thermoplastic	T
Moisture resistant thermoplastic	TW
Heat resistant rubber	R11
Moisture and heat resistant rubber	RHW
Moisture and heat resistant thermoplastic	THW

Table D–1

SIZES OF COPPER WIRE FOR SINGLE-PHASE, 115–120VOLT MOTORS AND A 2-PERCENT VOLTAGE DROP

Load (amps)	Minimum Allowable Wire Size: Wire in cable conduit, or earth: Types R,T,TW	Types RH, RHW, THW	Bare or covered wire overhead in the air[1]	Length of Wire to Motor (feet): Wire Size (AWG or MCM)[2] (Note: Compare the size shown below with the size shown in the column to the left of the double line and use the larger size.) 20	30	40	50	60	80	100	120	160	200	250	300	400	500
5	12	12	10	12	12	12	12	12	12	12	12	10	10	8	8	6	6
6	12	12	10	12	12	12	12	12	12	12	12	10	8	8	8	6	4
7	12	12	10	12	12	12	12	12	12	12	10	10	8	8	6	6	4
9	12	12	10	12	12	12	12	12	12	10	10	8	8	6	6	4	4
10	12	12	10	12	12	12	12	12	10	10	8	8	6	6	4	4	3
12	12	12	10	12	12	12	12	12	10	8	8	6	6	4	4	3	2
14	12	12	10	12	12	12	12	10	10	8	8	6	6	4	4	3	2
16	12	12	10	12	12	12	10	10	8	8	6	6	4	4	3	2	1
18	12	12	10	12	12	12	10	10	8	8	6	6	4	4	3	2	1
20	12	12	10	12	12	10	10	8	8	6	6	4	4	3	2	1	0
25	10	10	10	12	10	10	8	8	6	6	4	4	3	2	1	0	00
30	10	10	10	12	10	8	8	8	6	4	4	3	2	1	1	00	000
35	8	8	10	12	10	8	8	6	6	4	4	3	2	1	0	00	000
40	8	8	10	10	8	8	6	6	4	4	3	2	1	0	00	000	0000
50	6	6	10	10	8	6	6	4	4	3	2	1	0	00	000	0000	250
60	4	6	8	8	8	6	4	4	3	2	2	0	00	000	000	250	300
70	4	4	8	8	6	6	4	4	3	2	1	0	00	000	0000	300	350

Note: Use 125 percent of motor nameplate current for single motors.

[1] The wire size in overhead spans must be at least number 10 for spans up to 50 feet and number 8 for longer spans.

[2] AWG is American wire gauge and MCM is thousand-circular mil.

Table D–2
SIZES OF COPPER WIRE FOR SINGLE-PHASE, 230–240-VOLT MOTORS AND A 2 PERCENT VOLTAGE DROP

Load (amps)	Minimum Allowable Wire Size: Wire in cable conduit, or earth: Types R,T,TW	Types RH, RHW, THW	Bare or covered wire overhead in the air[1]	Length of Wire to Motor (feet): 20	30	40	50	60	80	100	120	160	200	250	300	400	500
				Wire Size (AWG or MCM)[2] (Note: Compare the size shown below with the size shown in the column to the left of the double line and use the larger size.)													
2	12	12	10	12	12	12	12	12	12	12	12	12	12	12	12	12	12
3	12	12	10	12	12	12	12	12	12	12	12	12	12	12	12	12	10
4	12	12	10	12	12	12	12	12	12	12	12	12	12	12	12	10	10
5	12	12	10	12	12	12	12	12	12	12	12	12	12	12	10	10	8
6	12	12	10	12	12	12	12	12	12	12	12	12	12	10	10	8	8
8	12	12	10	12	12	12	12	12	12	12	12	12	10	10	8	8	6
10	12	12	10	12	12	12	12	12	12	12	12	10	10	8	8	6	6
12	12	12	10	12	12	12	12	12	12	12	12	10	8	8	8	6	4
14	12	12	10	12	12	12	12	12	12	12	10	10	8	8	6	6	4
17	12	12	10	12	12	12	12	12	12	10	10	8	8	6	6	4	4
20	12	12	10	12	12	12	12	12	10	10	8	8	6	6	4	4	3
25	10	10	10	12	12	12	12	10	10	8	8	6	6	4	4	3	2
30	10	10	10	12	12	12	10	10	8	8	8	6	4	4	4	2	1
35	8	8	10	12	12	12	10	10	8	8	6	6	4	4	3	2	1
40	8	8	10	12	12	10	10	8	8	6	6	4	4	3	2	1	0
45	6	8	10	12	12	10	10	8	8	6	6	4	4	3	2	1	0
50	6	6	10	12	10	10	8	8	6	6	4	4	3	2	1	0	00
60	4	6	8	12	10	8	8	8	6	4	4	3	2	1	1	00	000
70	4	4	8	12	10	8	8	6	6	4	4	3	2	1	0	00	000
80	2	4	6	10	8	8	6	6	4	4	3	2	1	0	00	000	0000
100	1	3	6	10	8	6	6	4	4	3	2	1	0	00	000	0000	250

Note: Use 125 percent of motor nameplate current for single motors.
[1]The wire size in overhead spans must be at least number 10 for spans up to 50 feet and number 8 for longer spans.
[2]AWG is American wire gauge and MCM is thousand-circular mil.

Table D–3

SIZES OF COPPER WIRE FOR THREE-PHASE, 230–240-volt MOTORS AND A 2 PERCENT VOLTAGE DROP

Load (amps)	Minimum Allowable Wire Size: Wire in cable conduit, or earth — Types R,T,TW	Types RH, RHW, THW	Bare or covered wire overhead in the air[1]	Length of Wire to Motor (feet): 20	30	40	50	60	80	100	120	160	200	250	300	400	500
				Wire Size (AWG or MCM)[2] (Note: Compare the size shown below with the size shown in the column to the left of the double line and use the larger size.)													
2	12	12	10	12	12	12	12	12	12	12	12	12	12	12	12	12	12
3	12	12	10	12	12	12	12	12	12	12	12	12	12	12	12	12	12
4	12	12	10	12	12	12	12	12	12	12	12	12	12	12	12	12	10
5	12	12	10	12	12	12	12	12	12	12	12	12	12	12	12	10	10
6	12	12	10	12	12	12	12	12	12	12	12	12	12	12	10	10	8
8	12	12	10	12	12	12	12	12	12	12	12	12	12	10	10	8	8
10	12	12	10	12	12	12	12	12	12	12	12	12	10	10	8	8	6
12	12	12	10	12	12	12	12	12	12	12	12	10	10	8	8	6	6
15	12	12	10	12	12	12	12	12	12	12	10	10	8	8	6	6	4
20	12	12	10	12	12	12	12	12	12	10	10	8	8	6	6	4	4
25	10	10	10	12	12	12	12	12	10	10	8	8	6	6	4	4	3
30	10	10	10	12	12	12	12	10	10	8	8	6	6	4	4	3	2
35	8	8	10	12	12	12	10	10	8	8	8	6	4	4	4	2	1
40	8	8	10	12	12	12	10	10	8	8	6	6	4	4	3	2	1

Note: Use 125 percent of motor nameplate current for single motors.

[1] The wire size in overhead spans must be at least number 10 for spans up to 50 feet and number 8 for longer spans.

[2] AWG is American wire gauge and MCM is thousand-circular mil.

Table D–4
SIZES OF COPPER WIRE FOR THREE-PHASE, 230–240-VOLT MOTORS AND 2 PERCENT VOLTAGE DROP
(continued)

	Minimum Allowable Wire Size			Length of Wire to Motor (feet)													
Load (amps)	Wire in cable conduit, or earth: Types R,T,TW	Wire in cable conduit, or earth: Types RH, RHW, THW	Bare or covered wire overhead in the air[1]	20	30	40	50	60	80	100	120	160	200	250	300	400	500
				Wire Size (AWG or MCM)[2] (Note: Compare the size shown below with the size shown in the column to the left of the double line and use the larger size.)													
45	6	8	10	12	12	10	10	8	8	6	6	4	4	3	2	1	0
50	6	6	10	12	12	10	10	8	8	6	6	4	4	3	2	1	0
60	4	6	8	12	10	10	8	8	6	6	4	4	3	2	1	0	00
70	4	4	8	12	10	8	8	8	6	4	4	3	2	1	1	00	000
80	3	4	6	12	10	8	8	6	6	4	4	3	2	1	0	00	000
100	1	3	6	10	8	8	6	6	4	4	3	2	1	0	00	000	0000
120	0	1	4	10	8	6	6	4	4	3	2	1	0	00	000	0000	250
150	000	0	3	8	6	6	4	4	3	2	1	0	00	000	0000	250	300
180	0000	000	1	8	6	4	4	3	2	1	0	00	000	0000	250	300	400
210	250	0000	0	8	6	4	4	3	2	1	0	00	000	0000	250	350	500
240	300	250	00	6	4	4	3	2	1	0	00	000	0000	250	300	400	500

Note: Use 125 percent of motor nameplate current for single motors.

[1] The wire size in overhead spans must be at least number 10 for spans up to 50 feet and number 8 for longer spans.

[2] AWG is American wire gauge and MCM is thousand-circular mll.

BIBLIOGRAPHY

Air Conditioning and Refrigeration Institute. 1979. *Refrigeration and Air Conditioning.* Englewood Cliffs, N.J.: Prentice-Hall.

Alerich, W. N. 1975. *Electric Motor Control.* Albany, N.Y.: Delmar Publishers.

Althouse, A. D.; Turnquist, C.H.; and Brocciano, A. F. 1979. *Modern Refrigeration and Air Conditioning.* South Holland, Ill., Goodheart-Willcox Co.

Lang, V. Paul. 1979. *Principles of Air Conditioning.* Albany, N.Y.: Delmar Publishers.

Langley, B. C. 1974. *Electric Controls for Refrigeration and Air Conditioning.* Englewood Cliffs, N.J.: Prentice-Hall.

Mahoney, Edward F. 1980. *Electricity for Air Conditioning and Refrigeration Technicians.* Reston, Va.: Reston Publishing Co.

Patrick, D. R., and Fardo, S. W. 1979. *Industrial Electronics Systems.* Indianapolis, Ind.: Howard W. Sams & Co.

INDEX

INDEX